Environmental Information for Outer Continental Shelf Oil and Gas Decisions in Alaska

Committee to Review Alaskan
Outer Continental Shelf Environmental Information

Board on Environmental Studies and Toxicology

Board on Earth Sciences and Resources

Commisssion on Geosciences, Resources, and the Environment

NATIONAL RESEARCH COUNCIL
1994

NATIONAL ACADEMY PRESS   2101 Constitution Ave., N.W.   **Washington, D.C.** 20418

Library of Congress Catalog Card No. 94-65019
International Standard Book No. 0-309-05036-7

B-302

COMMITTEE TO REVIEW ALASKAN OUTER CONTINENTAL SHELF ENVIRONMENTAL INFORMATION

CHARLES G. GROAT *(Chair)*, Lousiana State University, Baton Rouge, Louisiana
JOHN J. AMORUSO, Amoruso Petroleum Company, Houston, Texas
JOHN C. CROWELL, University of California, Santa Barbara, California
F. RAINER ENGELHARDT, Marine Spill Response Corporation, Washington, D.C.
WILLIAM R. FREUDENBURG, University of Wisconsin, Madison, Wisconsin
KATHRYN J. FROST, Alaska Department of Fish and Game, Fairbanks, Alaska
CHRISTOPHER J.R. GARRETT, University of Victoria, Victoria, British Columbia, Canada
GEORGE L. HUNT, JR., University of California, Irvine, California
ROBERT R. JORDAN, Delaware Geological Survey, Newark, Delaware
STEPHEN J. LANGDON, University of Alaska, Anchorage, Alaska
JUNE LINDSTEDT-SIVA, Atlantic Richfield Company, Los Angeles, California
H. JOSEPH NIEBAUER, University of Alaska, Fairbanks, Alaska
JAMES OPALUCH, University of Rhode Island, Kingston, Rhode Island
ROBERT T. PAINE, University of Washington, Seattle, Washington
RICHARD M. PROCTER, PRAS Consultants, Calgary, Alberta, Canada
WILFORD F. WEEKS, University of Alaska, Fairbanks, Alaska
CLINTON D. WINANT, Scripps Institute of Oceanography, La Jolla, California

Staff
DAVID J. POLICANSKY, Project Director
WILLIAM E. BENSON, Sr. Staff Officer
TANIA WILLIAMS, Research Associate
KATE KELLY, Editor
RUTH CROSSGROVE, Information Specialist
ADRIÉNNE DAVIS, Sr. Project Assistant
SHIRLEY JONES, Project Assistant

Sponsor

U.S. Department of the Interior, Minerals Management Service *iv*

PREFACE

The favorable geological setting for vast accumulations of oil and gas in Alaska's coastal and offshore areas has attracted the attention of hydrocarbon resource explorers for several decades. The discovery of the giant Prudhoe Bay field in 1968, with an estimated ultimate recovery of 20 billion barrels of oil, assured a continuing interest in seeking and developing Alaska's hydrocarbon resources. Counterbalancing the great resource potential is a harsh physical environment and considerable concern about the impacts of exploration and development on the living resources, landscape, and livelihood and culture of the Native American population. Conflicting pressures from Alaska Natives, local and national environmental groups, and concerned individuals have had a strong influence on programs to explore for and develop oil resources on the Alaskan OCS and the Arctic National Wildlife Refuge.

Concern over the adequacy of environmental information to assess planned lease sales in the Chukchi Sea, Beaufort Sea, and Navarin Basin in 1991 led to language in the House appropriations report recommending that the Minerals Management Service (MMS) ask the National Research Council (NRC) to assess the adequacy of environmental information for leasing decisions. In constituting the committee to conduct the study, the NRC's Board on Environmental Studies and Toxicology (BEST) and the Board on Earth Sciences and Resources (BESR) selected individuals with expertise in environmental, social, and economic aspects of northern Alaska, as well as the geological and resource assessment fields important in determining whether the resource base that has attracted the intense interest in Alaska oil has been adequately characterized.

The inclusion of geologists and resource assessment experts in the fabric of the committee made the group a model of diversity. The strong sense of purpose and commitment to carrying out a thorough and impartial assessment shared by all committee members, combined with the mutual respect for each member's perspective, resulted in a remarkably congenial and productive atmosphere. Differences of opinion were discussed and resolved and egos did not get in the way of reasoned consensus. It was a mentally stimulating experience to work with these talented people and I thank them for the strong effort they put forth and the positive working atmosphere they created. David Policansky of BEST directed the staff effort for the project. His considerable experience in dealing with the committee culture, his understanding of the issues and information resources relative to the study, and his effectiveness in honing consensus were invaluable. Tania Williams of BEST efficiently carried most of the operational load and effectively handled reams of information that went into the study and comparable reams of draft text that emerged. William Benson of BESR provided both staff support and valuable insight into the geological aspects of the resource and environmental picture. The Director of BEST, James Reisa, provided useful guidance, especially during the formative stages of the study.

The study was carried out through meetings in Washington, at the NRC's Arnold and Mabel Beckman Center in Irvine, California, and in Barrow and Anchorage, Alaska. I would like to thank those who briefed the committee: Jeslie Kaleak, Benjamin Nageak, and Tom Albert with the North Slope Borough, Don Long, mayor of Barrow, Ronald Brower, president, and his staff at UIC/NARL, Burton Rexford, chair of the Alaska Eskimo Whaling Commission, Jacob Adams, president of the Arctic Slope Regional Corporation, and many other citizens of Barrow; ARCO and BP, who provided a tour of their North Slope production facilities; Paul Rusanowski of the Office of the Governor of Alaska; Ken Boyd, with the Alaska Department of Natural Resources; David Allen, Rosa Meehan, and Patrick Sousa with the U.S. Fish and Wildlife Service; Robert Spies and Art Wiener with the Exxon Valdez Oil Spill Trustees; Dorothy Smith and Pam Miller with Greenpeace; David van den Berg with the Northern Alaska Environmental Center; Pamela E. Miller with the Wilderness Society; and finally, MMS staff including Jerry Imm, chief of the Alaska region's Environmental Studies Branch, and Colleen Benner, our main contact with MMS for the study's duration. These people in Alaska, MMS staff members, state and federal agency personnel, oil and gas industry geologists, Alaska Natives and their organizations, and numerous other individuals provided reports and personal observations that constituted an important part of the information base for the committee's work.

The scientific community likes to believe that if good scientific information is available, good decisions will be made on that basis. Most of us know that other factors do, and in many cases should, enter into the resolution of resource development and environmental issues. Nevertheless, I am hopeful that the substantial information base that characterizes the valuable environmental, human, and energy resource base of the Alaska OCS and adjacent onshore areas will be used to its fullest extent in OCS decisions and that the gaps we have identified that need to be filled before new development takes place are filled expeditiously.

Charles G. Groat
Chair

CONTENTS

Executive Summary

A great deal of oil lies beneath Alaska's North Slope and the adjacent oceans. Some of this oil—approximately 12 billion barrels recovered to date—has been produced from the supergiant Prudhoe Bay field and nearby areas. However, despite evidence of large oil deposits beneath the outer continental shelf (OCS) off Alaska's North Slope, no oil has yet been produced from it. The Bering Sea off Alaska's west coast also has received attention from the oil industry, but no significant evidence of commercial deposits has been found there.

Producing oil from beneath the Chukchi, Beaufort, and Bering seas is difficult. The areas are remote from major industrial centers, transportation, and refinery facilities (except those close to Prudhoe Bay). The weather is harsh; ice covers the areas and threatens human structures for varying amounts of time during the year, and the area is dark for long periods in winter. Although engineering and technology have overcome many of these obstacles, increased awareness of the environmental and socioeconomic effects of oil and gas activities has resulted in increased environmental regulation of those activities and controversies over whether and how they should occur.

To help provide a scientific basis for evaluating these concerns, the U.S. House of Representatives, in its fiscal-year 1991 appropriations report for the Department of the Interior, requested that the Minerals Management Service (MMS) seek advice from the National Research Council (NRC) about the adequacy of scientific and technical information relevant to the potential environmental consequences of three Alaskan lease sales planned

1

for 1991 and 1992: Sale 126 (Chukchi Sea), Sale 124 (Beaufort Sea), and Sale 107 (Navarin Basin in the Bering Sea). MMS also asked the NRC to consider options other than conducting additional studies in case the information was inadequate in any respect. In response, the NRC convened the Committee to Review Alaskan Outer Continental Shelf Environmental Information, which has prepared this report.

The committee considered the adequacy of scientific information relevant to decisions concerning all phases of the OCS oil and gas process: leasing, exploration, development, production, and decommissioning. It also took seriously MMS's request for a review of options other than conducting additional studies. The committee was not asked and did not consider whether any oil and gas activities *should* be undertaken in Alaska.

It soon became clear to the committee that industry interest—and hence the potential for any effects of OCS activities—was much lower in the Navarin Basin than in the Chukchi and Beaufort seas. Indeed, after the committee began its work, Sale 107 was deferred for further review until 1996. Therefore, this report focuses much more on the Chukchi and Beaufort seas than it does on the Bering Sea.

To obtain information for its review, the committee read environmental impact statements for the three lease sales, technical reports obtained from MMS's Alaska Region; synthesis reports; peer-reviewed publications; several relevant NRC reports; and documents obtained from the North Slope Borough, the oil industry, and other organizations. In addition, the committee received briefings from MMS in Washington, D.C., and Anchorage, Alaska. In Alaska, the committee visited Barrow and Prudhoe Bay and adjacent oil fields. It received briefings in those places and held discussions with residents and elected officials of Barrow and the North Slope Borough and with officials of ARCO, British Petroleum, and Alaska Clean Seas. In Anchorage, the committee received briefings from state officials and representatives of environmental organizations, consulting firms, the University of Alaska, and others. It also received briefings from officials of ARCO, Mobil, and Shell at its meeting in Irvine, California.

The committee's major conclusions and its discussion of possible alternatives to additional studies are summarized below. Chapter 8 presents detailed recommendations for additional studies that would be useful for OCS decisions, particularly those concerning development, production, and decommissioning.

ADEQUACY OF INFORMATION
FOR OCS DECISIONS

The committee concluded that the environmental information currently available for the Chukchi, Navarin, and Beaufort OCS areas is generally adequate for leasing and exploration decisions, except with regard to effects on the human environment (i.e., socioeconomic effects, as defined in the OCS Lands Act). Prelease and lease-stage effects on the human environment—which can begin even before any physical or biological changes take place—are a special category of effects discussed in Chapter 6. In general, the information available for resource geology, the physical environment, biotic resources, spills, and mitigation and remediation activities adequately reflects the differences between Arctic OCS areas and other U.S. OCS areas where development and production have already occurred.

In making this determination, the committee recognized that OCS oil and gas activities present a variety of risks to the biological and human environment, and that even with sometimes sketchy knowledge, bounds could be put on the extent of those risks. Whether or not to accept the risks is a policy issue, not a scientific question. MMS's Environmental Studies Program (ESP) and oil and gas resources assessment efforts have yielded information that is credible and useful in establishing a general characterization of the living resources, physical conditions, social and economic setting, and likely oil and gas resources in the Arctic OCS. Although the geological characterizations are sound, the resource estimates appear conservative; this might affect any estimates of impacts and the ability to plan for the future.

In contrast, the committee concluded that the information is often not sufficient to support decisions about development, production, transportation, and siting of onshore facilities. Much of this information, of course, can only be obtained after exploration has identified a proposed development site. MMS should concentrate on fewer, longer-term studies of the living resources, social and economic conditions, and physical processes to develop the additional information needed. The studies' design and results should be peer-reviewed.

MAJOR INFORMATION GAPS

The major information gaps concern certain aspects of the potential social and economic impacts of OCS oil and gas activities, the extent and temporal distribution of ice gouging, and physical oceanography and its relation to important ecological processes. The greatest perceived and feared consequences of OCS development in the Arctic OCS concern oil spills and interference with marine mammals—especially bowhead whales—that are critical to the subsistence economy of the Alaskan Natives. These fears have been exacerbated by a lack of mutual trust among the parties involved and by limited public confidence in the motives of MMS, the oil industry, and those who oppose development. It is clear that substantial involvement of all potentially affected parties, including Alaska Natives, is a prerequisite for a successful approach to the development of Arctic OCS oil and gas resources. It is not clear to the committee whether any additional studies could provide enough information to satisfy all parties.

A major information need is to determine the extent and temporal distribution of ice gouging. Gouges in the seafloor caused by the grounding of large ice masses have been detected to water depths of more than 50 meters, and the gouges are abundant nearer the shore. However, there is insufficient information to determine the frequency of these events, which obviously can affect structures on the seafloor.

The model of ocean circulation that has been used is elaborate but inevitably inaccurate because of limited spatial resolution and major uncertainties about the physical processes and the mathematical conditions applied to model boundaries in the water. Nevertheless, the model's output is still a useful rough guide to the path and fate of spilled oil, although a simpler model might have been just as useful and credible. Another NRC committee recently reviewed the information available for oil and gas leasing in other OCS areas and concluded that trajectory estimates have relied too heavily on general circulation models (GCMs). This committee believes the same conclusion holds for the three lease areas involved here. With the relative lack of observations available, MMS's initial reliance on model predictions is understandable, but as previous NRC reviews noted, trajectory predictions must be tied more closely to field observations than they have been. For developing models, the committee concluded that the existing "bottom-up" approach of trying to produce the best possible model should be replaced with a "top down" or targeted approach driven by specific concerns about key ecological species and processes; this would give better focus to model development. For example, oceanographic models should be driven by a need to understand formation of leads (long, narrow passage of open-water passing through a region of sea ice that is navigable by surface vessels and air-breathing mammals) in spring, which in turn influence the migration and distribution of whales and birds. Modeling the fate of spilled oil in ice, however, is not very refined and requires improvement.

The MMS Environmental Studies Program (ESP) in Alaska is extensive, substantive, and high quality. It has established a credible characterization of social and economic conditions in Northern Alaska, and it has addressed several important questions. The program has carried out some studies that analyze potential changes associated with OCS development—particularly economic impacts. However, that work has failed to deal adequately with other issues that are critical to predicting and managing impacts on the human environment during all phases of the OCS development process, ranging from leasing-stage activities to the longer-term impacts of development and production. There is a particular need for attention to the social and cultural effects of leasing, exploration, development and production—including the gradual or long-term changes that can be expected to take place even in the absence of spills—as well as the broader range of sociocultural disruptions that can result from a spill and persist for years. As a result of omissions in the program to date, a significant fraction of the social and economic information that would be necessary for informed decisions about leasing, development, production, and termination is unavailable. As a corollary, not enough effort has been devoted to the pragmatic questions of what steps, if any, could be taken to avoid or lessen harmful consequences.

Finally, the committee is concerned that recent reductions in the budget and staff of MMS's ESP in Alaska—especially but not exclusively in social science disciplines—will limit the quality and quantity of future work.

ALTERNATIVES TO ADDITIONAL STUDIES

MMS's ESP is based on the sensible premise that if there is enough information, sound decisions can be made about balancing the negative and positive effects of development. Sometimes that is true, but in other cases it appears that additional studies are not cost-effective or even likely to resolve controversies over the effects of development, no matter how much time and money are spent on them. Because the committee was specifically asked to consider alternatives to additional studies, it has done so in many places in the report. Three examples are summarized here.

Bowhead Whales

One area of great controversy concerns the effects of OCS oil and gas development on the migration patterns of marine mammals, especially bowhead whales. Because hunting bowheads is so important to North Slope communities, there have been many studies of the effects of noise on bowhead behavior and migration. Despite those studies, the effects of noise are not resolved, and it is not clear whether further study can provide resolution. For this reason, the committee believes a reasonable solution is for MMS, the industry, and North Slope residents to attempt to reach agreement on the controversial matters—such as specific times and places that various activities occur—and how they should be adjusted, remedied, or mitigated in lieu of or concurrent with additional studies. Parties should not have to give up rights to other remedies as a precondition to begin the negotiation, although a mutual agreement would imply that the agreed-on course of action would be adhered to unless additional information showed to everyone's satisfaction that the course of action should be modified. If additional studies are conducted, they should be designed, implemented, and shared by all three parties.

There is no guarantee that this approach would be successful, but it seems unlikely that it could be less successful or more costly than the current system of dueling studies and reviews and their accompanying delays and recriminations. The approach might also work for many other controversial questions in Alaska and elsewhere (such as the effectiveness of causeway breaches for allowing fish migration). Several practitioners have developed considerable skill in this kind of dispute resolution.

It is of the utmost importance to build effective and mutually agreeable monitoring programs into any settlement. Monitoring is essential to evaluate an agreement's success and to provide a basis for facilitating similar agreements elsewhere.

Oil Spills

Although the largest spills are usually the result of marine transportation accidents, the frequency of such accidents has declined. Blowout frequency associated with petroleum drilling is 0.03% of wells drilled; the risk of "significant" oil loss—more than 50,000 barrels—is even more remote, at 0.01% of wells drilled. Even though large oil spills are rare, they are a major concern in the Arctic because of the perceived inability to clean them up or control them. It seems likely that no amount of additional study of biotic responses to oil, of the viscosity and flammability of oil under various conditions, or other laboratory studies or simulations of oil spills and cleanup technology will completely allay public concern about the effects of spills and the ability to clean them up or mitigate them. Experience with very large marine spills indicates that current cleanup technology and procedures need improvement and further study, especially in situ burning, which is the only current technique with the potential for recovering much of the oil from very large spills. Nonetheless, very few experimental spills have been performed in cold waters—or, indeed, in any OCS waters of the United States.

The very limited experience with cold-water spills does not allow scientists to predict confidently the effectiveness of cold-water spill counter-measures. There are few pre-spill data from habitats affected by accidental spills, and during a real spill, it has been difficult to set aside enough control or reference sites to detect the effects of the spill and the control and cleanup methods used. Experimental spills can be planned and controlled; pre-impact data can be collected to allow proper evaluation of spill and control and cleanup impacts.

In addition, the performance of spill-response teams cannot be optimum if they are never permitted to practice using real oil. Perhaps more important, the strengths and weaknesses of spill responses cannot be evaluated in the absence of practical experience, so sensible and appropriate protection, mitigation, and compensation plans cannot be designed.

Mitigating Long-Term Socioeconomic Effects

Among the long-term nonspill socioeconomic impacts that need to be dealt with are the potential for cultural erosion and for socioeconomic overadaptation (overadaptation refers to the difficulty of transferring specialized activities and facilities to other forms of economic production). The potential for overadaptation is exacerbated by the region's remoteness and the limited likelihood of successful economic diversification. Among the obvious possibilities for mitigating those foreseeable effects (as well as for helping to create more positive effects) would be the creation of trust funds. Working cooperatively with the state and affected local governments (including the Northwest Arctic Borough, as well as the North Slope Borough), MMS could explore the potential for mitigating longer-term effects through revenue sharing, as well as through other steps that could help mitigate the coming "bust" by building up locally controlled trust funds. If such funds were sufficiently large, they could even help cushion the impending end of the Prudhoe Bay revenues, with the principal being left intact and the annual proceeds being used to fund a significant fraction of the boroughs' current employment and other costs.

Environmental Information for Outer Continental Shelf Oil and Gas Decisions in Alaska

1 | INTRODUCTION

Although Alaska has magnificent scenery and bountiful natural resources, its remoteness and harsh climate can make it an expensive and difficult place to work, and much of Alaska's enormous oil and natural gas resources remain undeveloped. Indeed, until the discovery of the supergiant[1] Prudhoe Bay oil field in the late 1960s, Alaska's oil and gas were almost untapped. But the Prudhoe Bay field was so large—12 billion barrels (1 barrel contains 42 U.S. gallons) have been produced to date, and it has total recoverable reserves of 20 billion barrels—it justified the engineering feat of the Trans-Alaska Pipeline System (TAPS) to bring the oil to market. That pipeline, in turn, has made previously uneconomical discoveries potentially worth developing if the oil can be brought to the head of the pipeline in Prudhoe Bay. Nevertheless, the remoteness and great expense of producing oil from Alaska's outer continental shelf (OCS) requires that only large finds with reserves of perhaps hundreds of millions of barrels or more can be expected to be produced and delivered to the pipeline in the foreseeable future.

The potential for adverse impacts to the human and marine environment also needs to be factored into decisions with respect to oil development. The most visible and controversial impacts are those associated with accidental oil spills, despite the fact that offshore exploration and production account

---

[1] Although there is no hard and fast rule for categorizing oil-field size, the most used (and generally accepted) scale is small, <50 million barrels; large, 50-500 million barrels; giant, 500 million to 5 billion barrels; and supergiant, ≥5 billion barrels (Fitzgerald, 1980).

for only about 1% of oil released into the ocean each year (NRC, 1985; GESAMP, 1992). The 1969 oil spill in Santa Barbara, California, brought to public awareness the potential for harm caused by oil and gas exploration and drilling on the OCS. In the next two decades, two more spills—the 1979 *IXTOC I* blowout off Mexico's coast and the 1989 *Exxon Valdez* tanker accident in Alaska's Prince William Sound—reinforced the public perception that potential environmental damage results from spilled oil.

The effects of crude-oil spills in temperate marine environments are fairly well understood after 25 years of field and laboratory research. And even though the *Exxon Valdez* accident did not involve OCS oil, the magnitude of the spill (approximately 262 thousand barrels of North Slope crude oil) greatly influenced the national debate on the production of OCS oil and prompted even greater scrutiny of potential lease sales in Alaska. Whatever the actual impacts of spilled oil are, the emotional response to the possibility of spills is a significant factor in oil leasing and development considerations in Alaska.

THIS STUDY AND REPORT

In balancing the economic benefits of petroleum production against the potential for environmental damage, Congress and the Executive Branch can benefit from sound scientific assessments. Accordingly, the Minerals Management Service (MMS) of the Department of the Interior (DOI) requested this study as a result of fiscal year 1991's department appropriations report in which the U.S. House of Representatives recommended that MMS request the National Research Council (NRC) of the National Academy of Sciences to assess the adequacy of environmental information relevant to three Alaskan lease sales planned for 1991: Sale 126 (Chukchi Sea), Sale 124 (Beaufort Sea), and Sale 107 (Navarin Basin) (see Figure 1-1).

In response, the NRC's Board on Environmental Studies and Toxicology and its Board on Earth Sciences and Resources formed the Committee to Review Alaskan Outer Continental Shelf Environmental Information. The committee was made up of experts in geology, hydrocarbon resource assessment, oceanography, ice dynamics, spill response, ecology, biology, sociology, anthropology, and resource economics. It was to review information about the degree to which the sales in the three lease sale areas differed from those in OCS areas in which development and production have

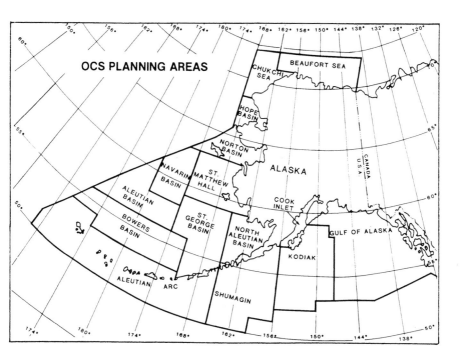

FIGURE 1-1   Outer continental shelf planning areas (Alaska). Source: MMS, 1987a.

already occurred.   It also was to review the degree to which information available on the three lease sale areas satisfactorily characterizes those differences.   The committee was asked to evaluate the approximate time required and the cost of obtaining additional environmental information and the likely improvement in decision-maker's ability to predict and manage environmental effects of OCS oil and gas activities that would result from having the additional environmental information.   Finally, the committee was asked to review options, other than conducting studies for obtaining additional environmental information—such as mitigation, lease sale stipulations, and acceptance of uncertainty (statement of task to the committee, September 1992).

Early in the committee's deliberations, discussions with industry representatives made it clear that industry's interest—and hence the potential for any effects of OCS oil and gas exploration and development—was much lower in the Navarin Basin (in the Bering Sea) than it was in the Chukchi and Beaufort seas. Indeed, in 1992, after the committee began its work, Sale 107 (Navarin Basin) was canceled and deferred for further review until 1996 (MMS, 1992a). Therefore, this report focuses more on the Chukchi and Beaufort seas than it does on the Bering Sea. The Bering Sea is not completely excluded, however, because it was in the committee's charge, because it has been studied extensively, because it has biological and physical influences on the Chukchi Sea and (to a lesser extent) on the Beaufort Sea, because ships supporting development in the Chukchi and Beaufort seas pass through the Bering Sea, and because it is possible that industry will reevaluate its interest in the area at some future time.

The committee recognizes that MMS's Environmental Studies Program (ESP) is not intended to be a broad, general science program like that of the National Science Foundation. Rather, it is a mission-oriented program, designed to answer questions about the environmental effects—including socioeconomic effects—of oil and gas exploration and production. Nonetheless, the answers to those questions must be based on sound science.

The committee used many resources in its work, including presentations from ESP staff; briefings by independent scientists familiar with the work supported by ESP; and discussions with Alaska Native leaders and with representatives of industry, environmental organizations, state government, and other interested parties at meetings in Barrow and Anchorage, Alaska. It toured facilities operated by ARCO, British Petroleum, and Alaska Clean Seas in the Prudhoe Bay area and was briefed by scientists and technical experts from those organizations. It received briefings from industry representatives at a meeting in Irvine, California. It also reviewed relevant scientific literature and documentation of the MMS planning and implementation process that leads to lease sales. The committee reviewed earlier NRC reports on the adequacy of environmental information for OCS decisions in California and Florida (NRC, 1989a) and in Georges Bank, Massachusetts (NRC, 1991a); reports of the Physical Oceanography Panel (NRC, 1990a), the Ecology Panel (NRC, 1992a), the Socioeconomics Panel (NRC, 1992b), and the final report of the full committee assessing ESP (NRC, 1993a); a report on the adequacy of the data base for petroleum hydrocarbon estimates of the Georges Bank area of the North Atlantic OCS (NRC, 1990b); and an evaluation of DOI's 1989 assessment procedures for undiscovered oil and gas resources (NRC, 1991b). It also reviewed the Environmental Impact Statements and Secretarial Issue Documents, geological reports, and other products of the ESP.

As background, the committee reviewed MMS documents that were available through 1992, other NRC reports, materials provided by industry and interest groups, as well as other materials available to professional scientists. MMS has informed the committee that the ESP continues to evolve, and recent requests for proposals confirm that. MMS officials have also indicated that they are taking into account recommendations made in the reports mentioned above.

Criteria for Adequacy

The committee concurs with previous NRC panels' operational definition of adequacy as it pertains to scientific information (NRC, 1989a, 1990a, 1991a, 1992a,b, 1993a). This definition, reviewed below, has two aspects: completeness and scientific quality.

Completeness

Because the body of scientific information grows continuously through research and discovery, completeness requires appropriate breadth and depth of basic scientific information in all relevant disciplines needed to illuminate the environmental risks associated with OCS oil and gas development. The criteria for completeness within disciplines for the three lease sales are described in the chapters that deal with the physical environment (Chapter 4), biotic resources (Chapter 5), and the human environment (Chapter 6).

Scientific Quality

The standards of scientific quality are repeatability, reliability, and validity of measurements and analyses, and they include the appropriateness of methods and subject. The working definition of scientific quality used by the committee was whether the methods described represent the current state of good practice in each scientific field; that is, whether the methods would be likely to pass peer review. That does not mean that actual publication in a peer-reviewed scientific journal is required, but rather that the quality of the data and of the scientific interpretations used for OCS leasing decisions should meet this basic scientific standard.

Application of the Standard

Although adequacy, or how much science is enough, can be defined for scientific purposes as outlined above, decisions must account for scientific uncertainty in the processes of assessing risks and in making predictions. How much uncertainty is acceptable for decision making is related to the state of the science, the perceived value of the resource or activity being considered, the nature of the risk, and public concern. The issue is how to balance the need to reduce uncertainty against the increased costs in time and money of doing the science required.

The process of determining how much science is sufficient for decision making about a lease should account for what kind of scientific knowledge will provide decision makers with an assessment of the potential environmental effects and risks—including the range of uncertainty—associated with oil and gas exploration and developments.

The definition of adequacy—and the specifics of adequacy in the case of the three relevant lease sales discussed in this report—does not address an ideal, but rather a minimum, essential amount of data or knowledge that is appropriate for informed decisions with respect to these lease-sale areas. The committee has evaluated only the adequacy of the scientific information as it provides the basis for informed decisions. It was not charged with evaluating the actual effects of OCS oil and gas development and production, and it has not done so here. The biological effects of specific OCS activities and the long-term effects of oil and gas development have been reviewed extensively elsewhere (for example, see NRC 1983a, 1985, 1989c; Boesch and Rabalais, 1987; Engelhardt, 1985a,b; COGLA, 1985a; GESAMP, 1992).

Report Organization

In this report, the committee focuses on what environmental information is needed to make oil and gas leasing decisions and if additional information is needed, determines the approximate time and financial cost involved in obtaining additional information. It also addresses how that information would improve the ability to predict and manage environmental effects of OCS oil and gas activities, and if applicable, determines the alternatives to obtaining that additional information. Each chapter of the report contains its specific conclusions and recommendations. In many cases, the committee was unable to estimate time and cost with confidence. Therefore, in those cases, estimates were not provided.

Chapter 2 describes operations. It reviews the physical effects of arctic industrial activity.

Chapter 3 considers the geological setting and the resource base for oil and natural gas. It reviews the general and petroleum geology of the areas in question, the adequacy of the geological data base, the adequacy and reliability of resource assessments, and engineering geology.

Chapter 4 considers the physical environment. It reviews the physical oceanic characteristics of the arctic region under consideration, and models of the region's circulation and oil spill trajectories.

Chapter 5 considers the biotic resources at risk, including marine mammals, birds, fish and fisheries, and benthic organisms.

Chapter 6 considers the social and economic environment. It reviews effects on the human environment, the distinctiveness of the high Arctic, and MMS's social and economic studies.

Chapter 7 reviews petroleum industry mitigation and remediation actions. It discusses the dynamics of working in an environment of sea ice, and it discusses the issues involved with cleanup of oil spills and the effectiveness of response measures.

Chapter 8 presents the committee's general conclusions and an alternative to additional studies regarding the adequacy of the environmental information that MMS uses to make OCS oil and gas exploration and development decisions.

Appendix A describes the environmental analyses mandated by Section 18 of the Outer Continental Shelf Lands Act of 1952 as amended in 1978. It outlines the federal decision-making process for establishing OCS lease sales.

DESCRIPTION OF THE AREA

In What Ways Are High Latitudes Different?

Although each area where oil and gas exploration and development has occurred or is planned has its share of unique environmental characteristics, the high-latitude planning areas considered here have relatively sensitive, remote, and hostile environments. The Arctic Research Commission (ARC, 1991) stated that "chronic environmental disturbances from both natural and human origins can create cumulative impacts on the arctic food chain; on gas reflux of greenhouse gases; and on the integrity of the permafrost, thus creating serious soil erosion and potential loss of biological diversity." Brown (1984) suggested that a smaller disturbance is required to produce an effect in the Arctic than is necessary in lower latitudes. Insults during the short and intensive growing season could produce effects of great amplitude, because many arctic species are near their tolerance limits for energy and nutrients and, therefore, tend to reproduce and grow slowly. This presumed fragility of the Arctic ecosystem is contested by some experts on the Arctic (Dunbar, 1985).

Arctic conditions are more hazardous than those found at most OCS drilling sites (Weeks and Weller, 1984). Sea ice, which often is present even during the summer in the Beaufort Sea, presents a major environmental challenge in high-latitude oil and gas exploration and development. Deformed pack ice forms pressure ridges that occur more frequently in the ice over the OCS than in ice further out to sea (Weeks and Weller, 1984). The keels of multiyear ice ridges can cause bottom scour that can damage pipelines and other exploration and production equipment. Also, the types of drilling platforms necessary to withstand the pressure of pack ice are costly and can take several years to build if work is restricted to the ice-free season. The tides and currents of the Arctic are not as severe as those in the Gulf of Alaska or the North Sea, but 3-m surges have been reported.

Rationale for Exploring for Oil in a Hostile Environment

The Alaskan OCS is one of the few areas left where industry believes very large oil fields—perhaps as large as 10-30 billion barrels—still can be found in U.S. territory. Before field exploration occurs, it is not possible to develop precise estimates of reserves in place. The uncertainty of these numbers should be recognized. It is worth noting that total domestic demand for refined petroleum products in 1992 was just over 17 million barrels per day (DOE, 1993), or about 6 billion barrels per year. Thus, the 12 billion barrels of oil produced from Prudhoe to date represents slightly less than 2 years' consumption at 1991 rates, or about 10% of domestic consumption for 20 years. Recoverable reserves of 10-30 billion barrels from the Alaskan OCS would supply all U.S. oil for 2-5 years at current consumption rates (or about 10% of domestic consumption for 20-50 years). Although arctic oil and gas are extremely expensive to produce and transport to market, the large potential reserves offer great economic incentive for exploration and security of supply from domestic sources are considerations for proceeding with exploration. The pressure to pursue OCS oil and gas development is spurred in part by declining production from the Prudhoe Bay field—ARCO estimates it will decrease 8-10% annually through the year 2000 (Davis and Pollock, 1992)—so new reserves that can be transported through TAPS are being sought.

TAPS has a finite economic and engineering lifetime, influenced by such factors as the price of oil, the amount of oil flowing through the pipeline, technology, general economic conditions, and global developments in energy technology. The General Accounting Office reviewed the Department of Energy's projections of the long-term viability for TAPS in April 1993. Its report concluded that it is not possible to pinpoint a year that TAPS will shut down. It is possible, however, to estimate a probable range of shutdown dates, 2001-2021. The variation in the range is due to uncertainty about TAPS minimum operating level, oil price, cost estimates, and estimates of discovered, but non-producing fields coming on line (GAO, 1993). Because it can take 10 years from the time of a discovery until oil flows in economic quantities, it is clear that there is a limited opportunity for proving reserves in the Alaskan OCS if oil is to be transported to market through TAPS. Minimum economic field sizes for Alaskan OCS waters have to be larger than those for any other offshore U.S. area, which imposes further limits to field development in this area.

In 1983, the DOI issued the Arctic Summary Report, which outlined three major factors that would determine whether oil and gas exploration could occur in the Arctic: the location and size of undiscovered oil and gas resources; future energy prices; and the policies developed by the federal government, the State of Alaska, and the North Slope Borough (Dugger, 1984). According to industry estimates, the "potential resources of Alaska onshore and offshore represent a very significant part of the total undiscovered resources left to be found in [the U.S.]" (Kumar, 1992).

Excellent source rocks have been found in the Chukchi Sea, and the Kuvlum discovery (perhaps 1 billion barrels) by ARCO proves that commercial quantities of oil are present in the Beaufort Sea. Because the costs of operating in the Arctic and of getting the oil to market are so high, the minimum economic field size is large and depends on the field's distance from the Prudhoe Bay TAPS terminal. In the Beaufort and Chukchi seas, the minimum economic size could be as large as 3 billion barrels of oil. Exploration of the Navarin Basin has not yielded promising results (Kumar, 1992).

Exploration and Production in the Lease Areas

Through 1990, the oil industry had spent $10 billion to drill 75 holes that were not commercially viable in themselves in the Alaskan OCS (Davis and Pollock, 1992). Between 1976 and 1990, 574 tracts had been leased and 21 wells had been drilled. The Beaufort Sea planning area has 469 active leases. In the Chukchi Sea planning area, 350 tracts have been leased and four wells have been drilled; all 350 leases are active. In the Navarin Basin planning area, 163 tracts have been leased and nine wells (including one deep stratigraphic test well) have been drilled; 22 leases are active (USGS, 1992a). As defined by MMS, a producible lease is one from which "oil, gas, or both [can be produced] in quantities sufficient to yield a return in excess of the costs, after completion of the well, of producing the hydrocarbons at the wellhead" (30 C.F.R. § 250.11 (1993)). Eight wells in the Beaufort Sea planning area have been classified as producible by MMS, although four have been relinquished. None of the eight wells can be considered economically productive at the time of the committee's deliberations. For instance, the outlook for activity in the Navarin Basin is not promising, but more exploratory drilling would be possible in the Chukchi Sea if partnerships between companies were formed. The search for major fields continues in the Beaufort Sea (Kumar, 1992).

MANAGEMENT OF OCS ACTIVITIES

Federal responsibility for development of mineral resources and conservation of OCS natural resources was established by the Outer Continental Shelf Lands Act (OCSLA) (67 State 462) of 1953, the Submerged Lands Act (67 State 29) of 1953, and the OCSLA amendments of 1978 (43 U.S.C. § 1331 et seq.) Leasing of OCS oil and gas resources is managed by MMS, which was formed in 1982 as a result of Secretary of the Interior James Watt's desire to consolidate responsibility for offshore oil and gas development within one agency. MMS includes some functions and personnel previously assigned to the Bureau of Land Management and the U.S. Geological Survey.

From 1954, when the oil and gas leasing program began, through 1991, the last year for which statistics have been published, federal OCS oil and gas development has provided almost 7.9% (9.2 million barrels) of total domestic oil production; about 14.7% (97.3 billion cubic feet) of domestic natural gas; and more than $96 billion in revenue from cash bonuses, lease rental payments, and royalties on produced oil and gas (MMS, 1992b). In 1991 alone, OCS production accounted for 11.7% of domestic oil, 24.2% of domestic natural gas, and almost $2.8 billion in revenue, of which $3.9 million was from Alaska (MMS, 1992b). From 1954 through 1991, there were 107 OCS lease sales, offering 165,697 tracts that included 903,029,994 acres. Only 13,039 (7.9%) of those tracts, which included 67,137,007 acres (7.4% of the acreage offered), were actually leased. In the Alaska region, 25,487 tracts were offered from 1954 through 1991; 1,566 (6.1%) of the tracts have been leased, an area of 8,617,658 acres (6.4%). Table 1-1 shows the regional breakdown of lease offerings, leases issued, and leases under supervision from 1954 through 1991.

The OCS lease sale schedule is established in accordance with a 5-year plan that sets forth the size, timing, and location of proposed leasing activities. The plan is developed in a 2-year process that includes consultation with coastal states and other federal agencies and an opportunity for public comment. Beginning in 1983, lease sales were offered for whole areas, instead of for selected tracts, to increase the numbers of blocks and leases and to encourage more drilling of exploratory wells in frontier areas (areas where there had been no oil and gas production), such as areas of deep water. The current plan, effective from mid-1992 to mid-1997, calls for 18 sales in 11 planning areas (MMS, 1992c). Since 1987, sales in several environmentally sensitive subareas have been deferred indefinitely (MMS, 1987a) and additional deferrals, cancellations, and leasing moratoria have occurred. Specific to this study, in 1992, the Navarin lease sale (107) was canceled and deferred for review until 1996 with the sale proposal for Norton, Navarin, and St. Matthew-Hall in the next 5-year program (1997-2001) (MMS, 1992c).

TABLE 1-1 Regional Summary, Federal Lease Offerings and Lease Status, 1954-1991

OCS Region	Lease Offerings	Offerings[1]		Leases Issued		Under Supervison[2]	
		Tracts	Acres	Tracts	Acres	Tracts	Acres
Alaska	15	24,587	135,557,952	1,565	8,617,658	865	4,741,240
Atlantic	8	9,160	22,033	410	2,334,089	56	318,818
Gulf of Mexico	72	130,063	706,209,315	10,593	53,645,232	6,475	32,633,593
Pacific	10	1,887	9,742,781	470	2,693,744	104	526,773
TOTAL	107[1]	165,697	851,532,081	13,309	67,137,007	7,500	38,220,424

[1] MMS has held two reoffering sales in offshore oil and gas lease areas: RS1 had 175 tracts (996,308 acres) in Alaska; RS2 had 140 tracts (785,091 acres) in Alaska, 155 tracts (882,444 acres) in the Mid-Atlantic, 232 tracts (1,320,819 acres) in the South Atlantic, and 27 tracts (153,716 acres) in northern and central California. These tracts and acres are included in the figures listed for offerings.

[2] All producing and nonproducing offshore mineral leases for which the Department of the Interior had responsibility as of December 31, 1991.

Source: Adapted from MMS, 1992b (Federal Offshore Statistics: 1991).

ESP Mandate

As it was amended in 1978, Section 18 of OCSLA requires MMS to manage the OCS oil and gas program with consideration for the economic, social, and environmental values of renewable and nonrenewable resources; for the marine, coastal, and human environments that could be affected; for the laws, goals, and policies of the affected states; and for the equitable sharing of developmental benefits and environmental risks among the various regions. The timing and location of leases must be selected, to the greatest extent practicable, to balance the potential for environmental damage, the potential for the discovery of oil and gas, and the potential for harm to the coastal zone (43 U.S.C. § 1344).

In requiring environmental studies, OCSLA establishes two goals for ESP. The first is to develop information needed for "the assessment and management of environmental impacts on the human, marine, and coastal environments of the OCS and the coastal areas that may be affected by oil and gas development" in a proposed leasing area (43 U.S.C. § 1346 (a)(1)). To the extent practicable, studies must be "designed to predict impacts on the marine biota, which may result from chronic low-level pollution or large spills associated with OCS production, from the introduction of drill cuttings and drill muds in the area, and from the laying of pipe to serve the offshore production area, and the impacts of development offshore on the affected coastal areas" (43 U.S.C. § 1346 (a)(3)). The second goal is to conduct additional studies subsequent to the leasing and development of an area or region as the secretary deems necessary and to "monitor the human, marine, and coastal environments of such an area or region to provide time-series and data-trend information which can be used for comparison with previously collected data to identify important changes in the quality and productivity of such environments, to establish trends in the areas studied and monitored, and to design experiments to identify the causes of changes" (43 U.S.C. § 1346 (b)). The many reviews and permits required at various stages of the OCS oil and gas process are shown in Figures 1-2 through 1-5 and are discussed in Appendix A.

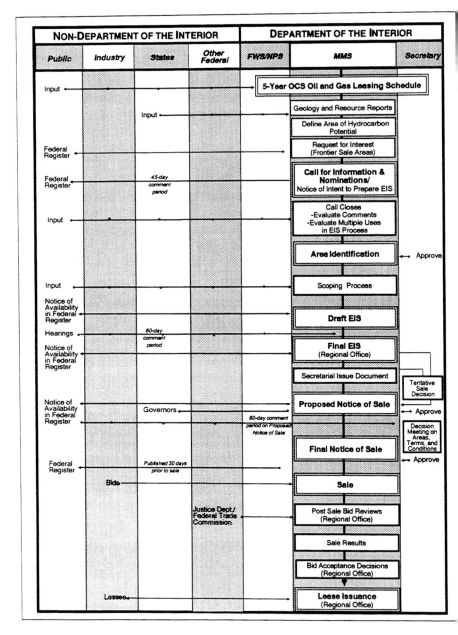

FIGURE 1-2   The pre-lease phase of the OCS oil and gas leasing process.
Source: MMS, 1991a.

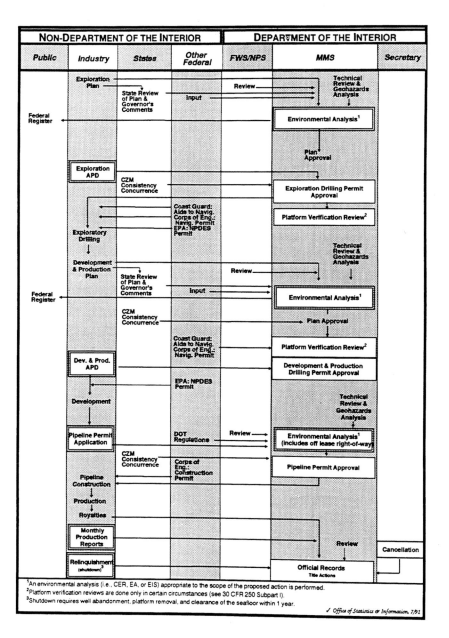

FIGURE 1-3 The post-lease phase of the OCS oil and gas leasing process.
Source: MMS, 1991a.

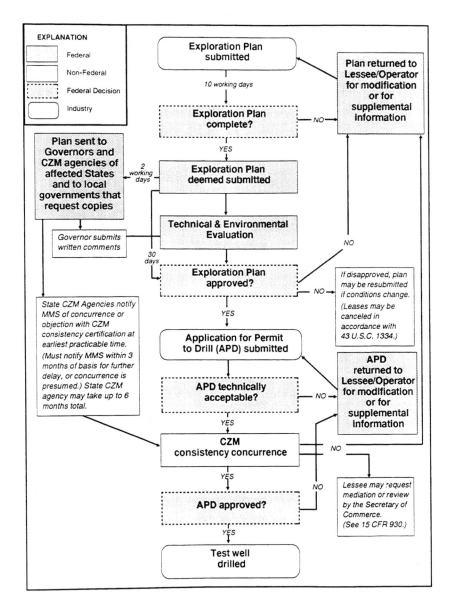

FIGURE 1-4   Exploration phase of post-lease activities. Source: MMS, 1991a.

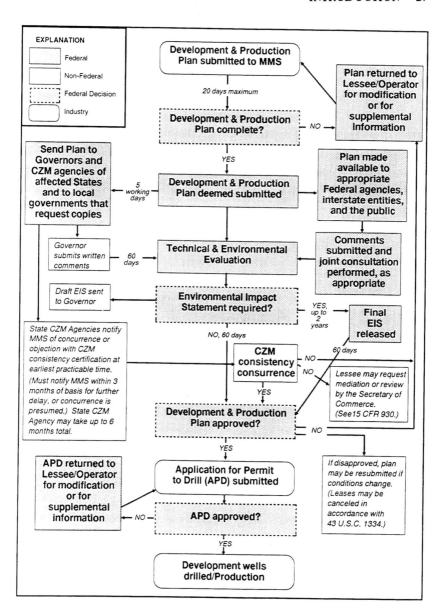

FIGURE 1-5   Development and production phase of post-lease activities.
Source: MMS, 1991a.

Environmental Information and
OCS Oil and Gas Leasing

The National Environmental Policy Act (NEPA), and judicial decisions in lawsuits over OCS leasing require that MMS must factually state what is known and not known about the environment for the decision maker in an EIS and address those subjects that have a bearing on the decision or its alternatives. Imperfect knowledge on a subject cannot stop a lease sale, but the decision maker must be aware of the inadequacy of that knowledge. This report is an attempt to focus on the merit of the existing scientific data base as developed in part by MMS's Environmental Studies Program and its adequacy for making leasing decisions. It is not a review of MMS's EISsor their adequacy.

A brief description of the decision-making process that the federal government uses to determine the timing, location, and buyers for an OCS lease is contained in Appendix A. Here we review the stages at which environmental information is used in that process.

The pre-lease phase of the OCS oil and gas leasing process is shown in Figure 1-2. The public has five opportunities to provide input. First, they can comment on the draft 5-Year OCS Oil and Gas Leasing Schedule. Second, they can provide comments to MMS after a Call for Information & Nominations is announced in the *Federal Register*. Third, after an area has been identified and the secretary approves it as a potential lease sale, the public can have input into the scoping process before a draft EIS. After a draft EIS is announced in the *Federal Register*, the public can participate in hearings or provide input during a 60-day comment period. Their final opportunity for input during the pre-lease phase occurs after a final EIS and Secretarial Issue Document are approved by the secretary of the DOI. A Proposed Notice of Sale is announced in the *Federal Register* along with an invitation to comment. Through the pre-lease phase, input is required from the state, industry, and other federal agencies.

The post-lease phase of the OCS oil and gas leasing process is shown in Figures 1-3 through 1-5. In this phase, public input is *not* required, although input *is* required from the state, industry, and other federal agencies. Appendix A discusses this in further detail.

After a lease has been issued and an exploration plan has been submitted, MMS conducts an Environmental Assessment (EA) as required by NEPA. If MMS finds that no significant impact on the environment will result, development and production can ensue.

An EIS may be required if the Development and Production Plan (DPP) submitted by a corporation lists seismic risks, areas of high ecological sensitivity, evidence of hazardous bottom conditions, or the use of new or unusual technology. Particular consideration is given to addressing significant adverse effects on the marine, coastal, or human environment that can result from the construction of new onshore and offshore facilities. Cumulative impacts are also considered. If the DPP is accepted, an Application for Permit to Drill is submitted. This application contains extensive information that allows MMS to evaluate the operational safety and pollution prevention measures of a proposed operation. If the application is approved, the lessee is given an Exploration Drilling Permit. The EA, including off-lease right-of-way, is required when the lessee submits a Pipeline Permit Application.

2 | DESCRIPTION OF OPERATIONS

PHYSICAL EFFECTS OF
ARCTIC INDUSTRIAL ACTIVITY

Activities required for the production of petroleum products in the Arctic include exploration, field development and production, and transportation. Their potential for causing harm to the arctic environment has been reviewed by Engelhardt (Engelhardt, 1985a,b; COGLA, 1985a) and more recently to the offshore in general by the International Maritime Organization's Group of Experts on the Scientific Aspects of Marine Pollution (GESAMP, 1992). The physical effects of industrial activity in polar environments can be outlined as follows:

• Seismic exploration requires the generation of sound waves, generated by mechanical devices and discharges from air guns. High-energy sound waves are propagated in unconsolidated sediments and consolidated rocks, as well as in seawater and biological material, and can cause local or regional disturbance of fish, marine mammals, and other wildlife. Variables that influence the severity of such damage include nearshore topography and ice cover.

• Vessel traffic, including icebreakers, can generate both noise and changes in ice breakup patterns, giving rise to concerns about disturbance of wildlife migration routes, especially the bowhead whale's.

• Exploration and production drilling generates noise, discharges of drilling wastes (cuttings, mud, drilling fluids) and production wastes (produced waters), and the risk of petroleum discharges through blowouts, all of which can cause mild to severe damage (mild to severe are expert judgments of the GESAMP committee and were not quantified).

• Oil spills draw attention in the Arctic as they do elsewhere. Spills in polar waters are not as well understood as are those in temperate waters, partly because there is less basic knowledge of the polar environment and partly because there is less specific knowledge about the effects of oil on polar organisms and ecosystems.

All phases of petroleum exploration and production can cause positive or negative changes in the arctic environment and in the lives of its people. Transportation of oil and oil products also results from petroleum development, and there is a potential for oil discharge in transportation accidents. The evaluation of effects is complex from industrial, environmental, and societal perspectives. This report concentrates on environmental information and its adequacy for evaluating potential impacts, including those generally perceived by many as beneficial, such as jobs and improved economy, health services, and schools, as well as those generally perceived as negative, such as social disruption, environmental damage, and the loss or erosion of traditional cultures.

Environmental Concerns

The possibility that oil spills can result from OCS oil and gas development and transport and the potential for damage to resources such as fisheries and to endangered species have caused great public concern. Other possible sources of harm associated with OCS development include the discharge of drilling muds and produced water at well sites. Seismic surveys and the construction and operation of platforms and pipelines can disturb wildlife and interfere with commercial, recreational, and subsistence fishing. Possibly adverse direct social and economic impacts are associated with the construction of onshore support facilities (NRC, 1978; MMS, 1987b). In Alaska, the possibility of damage to marine mammals that are important culturally and for subsistence, especially bowhead whales, is of particular concern, as is the possibility of other chronic environmental effects (Engelhardt, 1985b; GESAMP, 1992). The effects of oil spills and drilling-mud discharges were discussed in earlier NRC reports (NRC, 1975, 1983a, 1985). Table 2-1 summarizes some of the negative potential effects

TABLE 2-1   Negative Biophysical Concerns Associated with Industrial
Activities on the Coastal and Marine Environments in the Arctic

Activity Category	Issues
Seismic exploration	• Seismic energy releases • Vessel noise • Equipment noise (on ice)
Exploration and production drilling	• Liquid effluents • Solid wastes • Gaseous emissions • Operational noise • Vessel and vehicle (on ice) traffic • Aircraft traffic • Blow-outs
Hydrocarbon production and transport	• Produced water • Gaseous emissions • Operational noise • Vessel and/or vehicle traffic • Aircraft traffic • Ruptured storage units • Vessel noise • Tanker spills • Pipeline dredging • Pipeline rupture
Marine construction	• Artificial islands • Coastal bases • Channel excavation • High explosives • Causeways
Icebreaking	• Vessel noise • Traffic channels
Abandonment of production facilities	• Vessel and aircraft traffic • High explosives • Residual materials

of OCS exploration and development activities on the coastal and marine environments, but it is not exhaustive. Potential effects include pressure effects of some types of seismic surveys, disturbance of the seabed as a result of rig emplacement and platform installation, and noise at all post-lease phases of activity. Oil spills also are possible during the exploration stage, and there are potential effects from post-production activities such as the removal of platforms.

Even though OCS oil discharges are estimated at only 1% of oil spills into the world's oceans from all sources (NRC, 1985; GESAMP, 1992), they are a major source of public concern. From 1970 to 1991, there were 1,839 spills of more than 1 barrel each from OCS leases in the Gulf of Mexico—82 were more than 50 barrels—for a total estimated at 191,913 barrels. In the Pacific duing the same period, there were 34 spills—one of which was more than 50 barrels—for a total of 328 barrels. Total spillage for both regions from 1970 to 1991 thus was 192,288 barrels (MMS, 1992b). (Natural seepage in the Pacific OCS for the period is estimated to have been more than 28,000 barrels. Seeps release small amounts of oil gradually, rather than large amounts all at once. Consequently, weathering and reduction of toxicity are rapid, as are degradation processes. Spies (1983) found more animals near California seeps than in nonseep areas and suggested that the base of these food webs is made up of fast-growing populations of hydrocarbon-degrading microorganisms.) From 1964 to 1990, there were 24 oil spills of more than 1,000 barrels each from OCS leases for an estimated total of 447,678 barrels. The potential effects of offshore production-site abandonment pose unresolved concerns; there is little practical experience with post-production sites. Offshore production-site decommissioning is an area of study and experimentation. In the Gulf of Mexico and offshore California, platforms have been removed and recycled, as well as converted to reefs (MMS, 1987c, 1989, 1991a; GESAMP, 1992).

Exploration, development, production (Neff et al., 1987), and post-production activities can affect marine communities. Potential effects associated with each stage are distinct and require different suites of studies to predict their extent and duration. Exploration and production activities can result in the discharge of several classes of contaminants, including petroleum hydrocarbons (from drilling fluids, produced waters, and spillage) and trace metals (from drilling fluids and produced waters). The relative importance given to each contaminant class depends on previous exploration and production activities and on operational practices in a given area (Boehm, 1987).

Oil Spills

This assessment of the adequacy of information for leasing OCS areas under consideration for oil and gas exploration and development focuses on the marine environment of the Arctic. In general, the marine Arctic is characterized by three types of areas: those that are ice-free in summer; transition zones that contain deformed first-year and multiyear pack ice that show seasonal shifts in actual distribution and that are marked uniquely by the presence of leads and polynyas and a well developed under ice ecology; and the areas of permanent polar ice pack, which is predominantly multiyear and shows little biological activity. In a general way, the boundary between transition zone and permanent pack ice roughly corresponds to the edge of the Chukchi and Beaufort shelves.

Ice, especially where it is associated with areas of open water, forms a special habitat type that hosts a large number of species and ecological processes. The deformed pack ice at the ice front is an important environment for benthic and pelagic organisms (Dunbar, 1981). Epontic communities (assemblages living on the undersurfaces of the ice) are important to the productivity of arctic oceans. The open water of leads, polynyas, and pack ice is the seasonal habitat of many seabirds and marine mammals whose annual migrations tend to follow the lead patterns that form at breakup. The generally high productivity of these areas provides food for the birds, marine mammals, and fish.

An important environmental concern in the Arctic is the impact of oil spills. Ice greatly influences the fate of oil spilled in the Arctic, and the general behavior of oil in ice appears to be reasonably well understood, although the behavioral details are still debated and will undoubtedly vary with the properties of the oil and the details of the ice conditions (Mackay, 1985). Under-ice spills are retained in the irregular undersurface, but free oil can concentrate in breathing holes and in open leads and wind can herd oil against the ice edge. Low temperatures and restricted access to the atmosphere retard volatilization and prolong the toxicity of spilled oil. It is possible to hypothesize an increased vulnerability on the part of seabirds, marine mammals, and the ice-edge ecosystem in general. Biological timing—the time when an oil spill occurs in relation to the life histories or behaviors of populations or organisms—is a major factor in predictions of impact in these instances. Other environmental conditions need to be considered. The arctic habitat is divided into several parts. Clearly defined localities—such as shallow-water estuaries—form often exclusive and obligatory breeding or staging areas for many species. Walruses and spotted seals use traditional hauling-out areas. For many northern birds, the availability of appropriate onshore breeding areas is limited. Shallow coastal lagoons are used extensively by bird, mammal, and fish populations at specific times of the year, so an oil spill in one of these areas can have an impact on an entire species if it occurs at a time when a significant proportion of the population is present. Some of these areas are centers of subsistence harvesting by Alaska Natives (Iñupiat Eskimo) and others, and any impact can be significant both to the resource species and to the harvester. Experience with other spills—including the *Exxon Valdez*—tells us that the spills often create significant, longer-term socioeconomic disruption with consequences in turn for the credibility of industry and government.

One area of the arctic environment that can show little or no post-spill biological impact is the intertidal community. Although intertidal areas are productive and highly diversified near temperate oceans, they are of less year-round significance in the Arctic because of the biologically limiting effects of seasonal ice and low temperature. If an oil spill were to occur during a foraging or staging season for large numbers of shorebirds, however, there would be a risk of harm to those animals.

Routine Operations and Habitat Disturbance

There is direct disturbance of the seafloor during platform construction. This is limited to a very small area. A change in benthic habitat also can occur from the discharge of drill cuttings. Most damage, however, is caused when oil-based drilling muds are used (Davies et al., 1989; NRC, 1985), although they are not used in Alaskan waters. This kind of change is greater with production than with exploration drilling, simply because the amount of material discharged in a single location is much greater during production. Some contamination has been found in sediment and benthic organisms several kilometers distant from the source site (GESAMP, 1992). From a larger geographic perspective, the habitat changes are localized and recovery by recolonization appears to be possible.

Waste Discharges

Exploration involves the drilling of wells to determine the nature of potential gas and oil reservoirs in the aftermath of regional geological research and geophysical analysis. Exploration generally lasts only weeks to a few months, and usually involves just one well. It is important to note that it can take up to 10 years from permit application to exploration. Therefore, impacts are not concentrated in time. Exploration-well discharges are composed mostly of drilling fluids and drill cuttings. Very localized effects, such as the smothering of bottom dwellers, can be seen in the immediate vicinity of a platform.

Development drilling follows the drilling of successful wildcat or exploratory wells and can continue into the production phase of operations when the commercial reservoir is being produced. These operations can involve a large number of wells, and they are typically conducted from a fixed platform. There is, however, a trend in offshore development areas to use many deviated wells from a single platform with subsea production equipment to develop a field, although these have not been used at ice-covered sites. Discharges include drilling fluids, drill cuttings, and well treatment fluids. Very localized effects can be seen here as well.

Production activities begin as each well is completed during the development phase. The production phase involves active recovery of oil or gas from producing formations. Development and production activities can occur simultaneously until all wells are completed or reworked. Production water waste streams are the most significant discharges during production operations. The current state of knowledge regarding environmental effects and discharge controls was presented in the proceedings of a recent conference on the topic (Ray and Engelhardt, 1992). The bulk volumes of typical discharges from offshore oil and gas activities are shown in Table 2-2.

The largest volume of material discharged from production activities is formation water derived from the petroleum reservoir. Drilling fluids and cuttings, ballast water, and storage displacement water are also discharged

TABLE 2-2   Typical Quantities of Wastes Discharged During Offshore Oil and Gas Exploration and Production Activities

	Approximate amounts (tons)
Exploration well	
Drilling mud	
Periodically	15 - 30
Bulk at end	150 - 400
Cuttings (dry mass)	200 - 1,000
Base oil on cuttings	30 - 120[a]
Production site	
Drilling mud	45,000[b]
Cuttings	50,000[b]
Production water	1,500/day[c]—varies greatly with reservoir

[a] Actual loss to environment may be higher (Chénard et al., 1989).
[b] Estimate based on 50 wells drilled from a single offshore production platform, drilled over 4 to 20 years (Neff et al., 1987).
[c] Single platform (Menzie, 1982).
Source: GESAMP, 1992.

but are largely reinjected. Minor discharges can include produced sand, deck drainage, well completion and workover fluids, cement residues, blowout preventer fluid, sanitary and domestic wastes, gas- and oil-processing wastes, cooling water, desalination brine, and test water from fire-control systems.

Environmental Effects

The biophysical effects of waste discharges from ocean platforms have been reviewed by GESAMP (1992):

• Data from the North Sea and the Gulf of Mexico show that changes can occur in benthic communities close to production sites. The changes are attributed mainly to the discharge of drilling wastes, including cuttings (NRC, 1985).

• Changes in the benthic communities around production sites can be demarcated into zones of effect, and the extent of these zones depends on the amount and type of industrial activity and on the physical oceanographic setting. From a regional perspective, the total area of the seabed affected is very small or negligible.

• Produced water discharges can affect benthos, but are unlikely to be significant except in relatively shallow water areas, perhaps less than 20 to 30 m.

• The effects of drilling discharges from single-well exploration and multiple-well development and production activities are similar qualitatively, but differ greatly in magnitude, spatial extent, and predicted recovery rates.

• Current evidence indicates that the recovery of affected sites begins soon after drilling ceases.

• It is highly unlikely that the discharge of chemical wastes from offshore exploration and production causes any hazard to human health.

TABLE 2-3   Major Permitted Discharges and Potential Impact-Causing Agents Associated with Offshore Oil and Gas Exploration and Production in the United States

Drill cuttings	1,100 tons/exploration well, less for development well
Drilling fluids	900 tons/exploration well, 25% less for development well
Cooling water deck drainage, ballast water	May be treated in an oil/water separator
Domestic sewage	Primary activated sludge treatment
Sacrificial anodes, corrosion, antifouling paints	May release small amounts of several metals (Al, Cu, Hg, In, Sn, Zn)
Production water	Treated in oil/water separator to reduce total
Hydrocarbons	To mean of 48 ppm, daily maximum 72 ppm

Source: Neff et al., 1987.

• Although the discharge of production water from platforms can taint fish on the basis of absolute concentration, dilution of the plume within 1,000 m of the discharge site renders the risk minor.

Although the GESAMP (1992) review suggested that the effects of waste discharges are limited, this conclusion was based predominantly on data from temperate environments. The GESAMP (1992) evaluation did make a recommendation relevant to this evaluation of MMS Alaskan OCS studies: It said that the information is drawn almost exclusively from experience in North Sea and North American offshore operations, and that additional information should be obtained from a larger variety of environments and from latitudes in more vulnerable localities, such as shallow or enclosed waters and the Arctic.

Regulatory Controls

Regulatory controls that govern the discharges from offshore oil and gas drilling vary among jurisdictions and change over time as new information becomes available on the effects of discharges. In the United States, offshore discharges are regulated through a permitting system administered by the U.S. Environmental Protection Agency. Major permitted discharges and agents that can cause damage are summarized in Table 2-3 (Neff et al., 1987).

3 GEOLOGICAL SETTING AND HYDROCARBON Resource Base

INTRODUCTION

The coastal region of the North Slope of Alaska produces more petroleum than any other area of the United States. The supergiant[1] Prudhoe Bay field, many times larger than any other U.S. oil field, is one of five reservoirs that produce oil (Table 3-1).
Although the Prudhoe field tends to overshadow the others, Kuparuk is a giant field in its own right, much larger than fields in the lower 48 states, and "little" Milne Point would rank as a major field in Oklahoma or Texas.

GENERAL GEOLOGY OF THE BEAUFORT AND CHUKCHI OUTER CONTINENTAL SHELVES

The outer continental shelf (OCS) north of Alaska stretches for more than 1,000 km from the border with Canada on the east to the International Date Line on the west. Along the margin of the Beaufort Sea, its width from the shoreline to the slope break at a water depth of 200 m ranges from 70 to 160 km. West of Point Barrow it widens into the broad Chukchi shelf, a triangular region about 500 km on a side. The region is therefore comparable in area to the Gulf of Mexico offshore region from Pensacola, Florida, westward to the Mexican border.

---

[1]A "giant" field is usually considered as $\geq 500 \times 10^6$ bbl. "Supergiant" has been used differently by different writers, but all agree on $\geq 500 \times 10^9$ bbl.

TABLE 3-1 Producing Fields of the Prudhoe Bay Area

Field	Estimated Total Reserves	Estimated Recoverable
Prudhoe Bay	22 Bbbl[a]	12 + Bbbl
Kuparuk	6.4 Bbbl	2-3 Bbbl
Endicott	1 Bbbl	400 Million bbl
Lisburne	2 Bbbl	200 Million bbl
Milne Point[b]	500 ? Million bbl	100 Million bbl

[a] bl = barrel; bbl = barrels; Bbbl = billion barrels (10^9 barrels)
[b] Milne Point is not currently producing. It may be reactivated in the near future.
Source: BP Exploration (Alaska) Inc., 1991.

The shelf is the geological extension of the North Slope of Alaska and is underlain by a variety of sedimentary strata, including many formations that constitute source rocks and reservoir rocks for petroleum and natural gas. These range in age from Silurian to Tertiary (see Figure 3-1 for the generalized geological time scale) and are arranged in complex stratigraphic packages that document several major Earth events (Hubbard et al., 1987; Grantz et al., 1990; Haimila et al., 1990), including mountain building, uplift, and deep erosion; crustal stretching that formed deep basins that captured organic-rich sediments; and overprinting deformations that resulted in many types of geological structures that could be oil and gas traps.

The Department of the Interior's Minerals Management Service (MMS) report on the Chukchi Sea Planning Area (MMS, 1987d) summarizes the stratigraphy as follows:

> Three major stratigraphic sequences are recognized in north-western Alaska. (1) the Franklinian sequence (Precambrian to Middle Devonian metasedimentary rocks), which comprises the acoustic and economic basement complex throughout most of northern Alaska; (2) the Ellesmerian sequence (Devonian to early [sic] Cretaceous), which is composed of northerly sourced clastic and carbonate rocks; and (3) the Brookian sequence (early [sic] Cretaceous to Tertiary), which comprises a clastic

Era	Period	Epoch	Million Years Ago
CENOZOIC	Quaternary	Holocene (Recent)	
			0.01
		Pleistocene (Glacial)	
			1.8–2.8.
		Pliocene	
			4.9–5.3
		Miocene	
			23–26
	Tertiary	Oligocene	
			34–38
		Eocene	
			54–56
		Paleocene	
			63–66
MESOZOIC	Cretaceous		
			135–141
	Jurassic		
			200–215
	Triassic		
			240
PALEOZOIC	Permian		
			290–305
	Pennsylvanian (Upper Carboniferous)		
			330
	Mississippian (Lower Carboniferous)		
			360–365
	Devonian		
			405–415
	Silurian		
			435–440
	Ordovician		
			495–510
	Cambrian		
			570

PRECAMBRIAN

FIGURE 3-1   Generalized geological time scale.

wedge that was shed generally northward away from the Brooks Range orogen. The Ellesmerian sequence is separated from the underlying basement by the Ellesmerian unconformity (EU) and from the overlying Brookian sequence by the Lower Cretaceous unconformity (LCU). Offshore, the Ellesmerian seismic sequence is subdivided into lower and upper parts by a Permian unconformity (PU). The Brookian seismic sequence is subdivided into lower and upper parts by the mid-Brookian unconformity (mBU) of Late Cretaceous to Tertiary age.

The generalized stratigraphic sequence is illustrated in Figure 3-2; that of the National Petroleum Reserve is shown[2] in Figure 3-3. The structures that are known to be oil-productive in the Prudhoe Bay region extend northward offshore and to the west and northwest. The oil industry studies the subsurface geology of this vast region and sorts out the region's complexities to locate exploratory wells. The area's complexities suggest that many test wells will be needed before the amount of oil and gas there can be estimated accurately. Known onshore occurrences, however, make it likely that the offshore region contains significant quantities of oil and gas that can be developed if the economic obstacles due to the operating problems in the hostile arctic environment can be overcome. Indeed, the number and vertical distribution of oil-bearing strata in the Prudhoe Bay region (Figures 3-4a and 3-4b) and the probable extension of many of these formations suggest that the North Slope of Alaska and the adjacent continental shelf contain large reserves and that this region will continue to be explored and developed for some time. The similarity of the geology of the North Slope to several other passive continental margins adds to our confidence in this interpretation.

PETROLEUM GEOLOGY

Resource estimates of the offshore areas of Alaska emphasize a

───────────────────

[2]The National Petroleum Reserve-Alaska is a large area of the north coastal area of Alaska extending from Point Belcher on the west to the margin of the Prudhoe Bay area on the east.

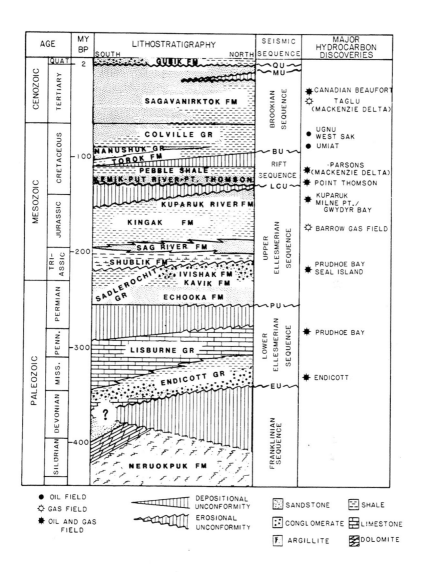

FIGURE 3-2   Generalized lithostratigraphic column showing the relationship of onshore rock units in northern Alaska to offshore seismic sequences in the Beaufort Sea Planning Area. Significant hydrocarbon discoveries in northern Alaska and Canada are shown by their reservoir formations. Source: Craig et. al, 1985.

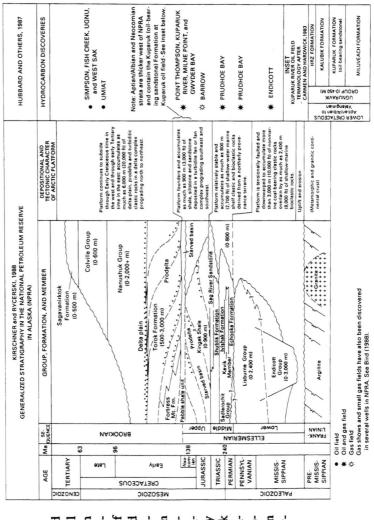

FIGURE 3-3 Generalized stratigraphy in National Petroleum Reserve in Alaska, showing interpreted environments of deposition for selected units, stratigraphic position of hydrocarbon discoveries, and Kuparuk River oil field terminology proposed by Carman and Hardwick (1983). Source: Haimila et al., 1990. Reprinted with permission of the authors; copyright, 1990.

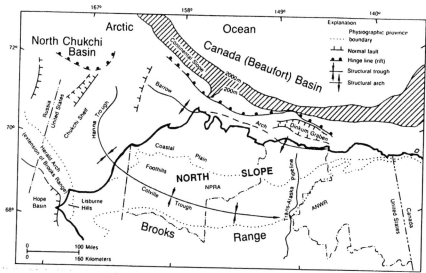

FIGURE 3-4a   Generalized structural features and geological framework of
onshore and offshore northern Alaska. Source: Adapted from Thomas
et al., 1991.

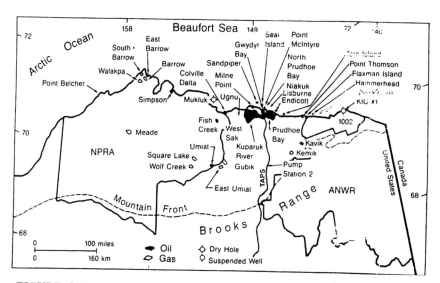

FIGURE 3-4b   Known oil and gas accumulations, selected dry holes and
suspended wells, (and NPRA-ANWR boundaries), North Slope,
Alaska. Source: Thomas et al., 1991.

substantial energy endowment in the region. The estimates are, of course, widely variable depending on what method is used and the amount and quality of geological information—but all current estimates generally agree that about 25% of the country's undiscovered oil resources are offshore Alaska. This potentially huge energy endowment has attracted drilling in most of the offshore Alaska basins on leases previously obtained from the federal government. The initial exploratory drilling programs also have led to consideration of additional lease sales, the subject of this report: Navarin Basin, Sale 107[3]; Chukchi Sea, Sale 126; and Beaufort Sea, Sale 124.

None of the offshore explorations had revealed oil in quantities large enough for commercial production until October 1992, when ARCO announced an apparently commercial discovery in its No. 1 Kuvlum well in the eastern Beaufort Sea, about 97 km east of Prudhoe Bay and 26 km offshore on federal acreage. Although only the discovery well has been drilled to date, many industry sources believe that the new field could contain reserves of 1 billion barrels or more. The real extent of the field and the number of barrels in it will be determined by confirmation and delineation wells. Meanwhile, regardless of the absolute field size, the ARCO discovery highlights the reserve potential of offshore Alaska.

Chukchi Sea

The Chukchi Sea province, approximately 400 km long and 240 km wide, is off the coast of Alaska between Barrow and Cape Lisburne. The province is limited on the north by the Chukchi Borderland, considered too far north and in waters that are too deep to be drilled feasibly in the near future (Thomas et al., 1991). The province is bordered on the east by the edge of the Arctic Platform and on the west by the margin of the Chukchi Platform. To the south and southeast the province is bordered by the Herald Arch and the seaward extension of the Brooks Range (MMS, 1989).

The Chukchi Sea province is composed of three primary sedimentary basins: the Colville Basin, the Central Chukchi Basin, and the North Chukchi Basin. The Colville Basin contains about 6,100 m of Brookian sedi-

---

[3] Although MMS canceled and deferred the Navarin Sale for further review until 1996, a short description of the area is included in this report.

ments in a northwest-to-southeast-trending foredeep trough north of the Brooks Range orogen. The Central Chukchi Basin is a north-trending offshore extension of the Arctic Alaska Basin. It contains 12,200 m of Ellesmerian and Brookian sediments. The North Chukchi Basin contains more than 13,715 m of lower and upper Brookian clastics (MMS, 1989, 1991a). The stratigraphic sequences appear to be offshore extensions of the western North Slope stratigraphy. As at Prudhoe, the underlying Franklinian sequence of metamorphosed Cambrian to Middle Devonian rocks are the lower limits of possible petroleum deposits in the Chukchi province. The Ellesmerian and Brookian sections provide the source rocks and reservoir potential for petroleum in the area. The rocks of the Ellesmerian section (late Devonian to early Cretaceous in age) are shallow-water limestones and marine and nonmarine sandstones and shales derived from a northerly source. No COST (continental offshore stratigraphic test) well has been drilled in the province to provide public information about the likely sources and reservoirs of the Ellesmerian section, but seismic data indicate that the source of the offshore section was from the west (MMS, 1989). The Cretaceous to Tertiary Brookian section is composed, onshore and offshore, of deltaic sediments derived from the Brooks Range to the south.

Geochemical studies from onshore areas show that the Ellesmerian and Brookian sections contain source rocks that could provide significant quantities of oil to potential reservoir beds, most of which seem to be in the Lisburne and Sadlerochit groups of the Ellesmerian section. They produce prolifically onshore and are believed to have similar capabilities at various places offshore. The shallower Brookian sequence, which is productive onshore and in the Canadian Beaufort, is also expected to have good offshore potential.

Tens of thousands of kilometers of high-quality seismic surveys with grid spacings as close as 1.6 × 1.6 km (1 mile × 1 mile) in some areas have been obtained in the province. MMS, for example, had acquired 73,000 miles of common depth point (CDP) seismic profiles over the area by the end of 1987. This extensive net of seismic control enabled the identification of many structural and stratigraphic traps. Structural traps are common throughout the area and could predominate in the Central Chukchi and North Chukchi basins. Both fold- and fault-related structural traps are present; some of these are complicated by unconformities (erosional gaps) in the Ellesmerian and Brookian sections. Stratigraphic traps could predominate along the flanks of the Arctic and Chukchi platforms. The first federal lease sales were conducted in the Chukchi province in 1988. Bids for OCS Sales 97 and 109 totaled $18 and $478 million, respectively. From 1989 through 1991, Shell Oil drilled five wells on structural prospects in the Colville and Central Chukchi basins. Detailed descriptions of the wells, with the exception of the Klondike No. 1 drilled in 1989, have not been released, but there is no report that any of the wells found a commercial quantity of oil or gas.

General statements from the industry indicate that, although the wells did not contain commercial accumulations, some oil is present and is effectively trapped in the structures. Source rocks were identified; particularly the Shublik Formation of the Ellesmerian sequence. High-quality reservoir rocks were not as widespread as had been hoped. However, the five initial test wells provide only a very sparse sampling of the stratigraphy of the province, and subsequent wells located with more emphasis on stratigraphy could identify potential reservoir rocks.

Results of the first drilling phase are still being evaluated, but they appear to provide encouragement for further exploration of the province. The key elements of oil accumulations have been confirmed and only the finding of better quality reservoir rock remains for discovering commercial accumulations.

Operations in the severe offshore arctic environment are difficult and operational flexibility is necessary for a successful drilling program. Costs are very high and cost-sharing industry partnerships might be necessary. The 1991 OCS Sale 126, in which only $7 million was bid by industry, reflects the difficulties of drilling in the Chukchi province. It also reflects depressed oil prices and opposition to OCS exploration.

Nevertheless, all the elements for oil accumulations have been confirmed in the Chukchi province. The next phase of exploratory drilling might find the 1-billion-barrel field, considered to be the minimum economic field size that could establish the Chukchi as a prolific oil province.

Beaufort Sea

The Beaufort shelf is in the offshore area of Alaska between the Canadian border on the east and Barrow on the west, generally parallel to the coast. It is approximately 560 km long and 96 km wide. It lies off-shore of the Arctic National Wildlife Refuge (ANWR) to the east and the National Petroleum Reserve, Alaska (NPRA), to the west. The area of interest is bounded on the south by the Barrow Arch and on the north by a shelf edge and slope into the deep Canada Basin (Figure 3-4a) (Thomas et al., 1991). The shelf can be divided into two major provinces, separated by a highly faulted Hinge Line (MMS, 1989): The Arctic Platform lies south of the Hinge Line and the Brookian Basin lies north (MMS, 1990). The northern limit for prospecting in the Brookian Basin is the deep water of the Beaufort Basin (Thomas et al., 1991).

The province can be divided into six primary centers of sediment deposition: Northeast Chukchi Basin, Nuwuk Basin, Arctic Platform, Arctic Basin, Dinkum Graben, and Kaktovik Basin. The Franklinian crystalline metamorphic rocks are the seismic and economic basement of the province. Overlying this basement is the Ellesmerian sequence, which provides the source and reservoir rocks of the Prudhoe Bay and other North Slope fields.

The Northeast Chukchi Basin is structurally isolated from the better known Ellesmerian sequence to the east below the Beaufort Sea, and the rocks in this basin are so far unproductive. They are considered a separate sequence. The Ellesmerian sequence, derived from a northern source, thins as it laps onto the ancestral landmass to the north. "Thick wedge-shaped layers of Ellesmerian sequence rocks are present in the southern part of the province; they become thinner to the north and are absent entirely in the northern part" (MMS, 1990). The sequence is generally present on the Arctic Platform and the Nuwuk and Kaktovik basins, but is missing to the north in the Arctic Basin. Because the Ellesmerian section produces most of the oil in the North Slope fields, the areas where these sediments are present are considered to be the most promising for oil exploration. Uplift of the Arctic Platform in early Cretaceous time terminated deposition of the Ellesmerian sediments; and erosion of the youngest Ellesmerian strata produced an unconformity that truncated the section.

Rift zone grabens (fault basins), such as the Dinkum Graben, developed within the Arctic Platform. The rifts were filled with a thick Lower Cretaceous clastic section, known as the rift sequence. By the late Cretaceous, Brookian sediments, derived from the Brooks Range to the south, were deposited over the Arctic Platform, into the Nuwuk and Kaktovik basins, and over the Hinge Line and shelf edge to a thickness of 12,200 m in the Arctic Basin to the north.

The province has been covered by tens of thousands of kilometers of high-quality seismic surveys, with grid spacing as close as 0.5 × 0.5 mile in some portions.   By 1987, for example, MMS had acquired 49,000 line miles of CDP seismic data over the province (MMS, 1989).   As a result of this seismic coverage, many structural and stratigraphic features that could contain oil have been identified.   These include anticlinal folds, fault traps, unconformity traps, stratigraphic traps, and their combinations.   Excellent source rocks have been identified in the Ellesmerian section of wells in Alaskan state waters and onshore.   Prudhoe Bay Field, the largest oil field in North America, and its surrounding fields left no doubt that oil was generated and entrapped in the province.   Excellent-quality reservoir rocks that are present onshore are believed to be present offshore as well.   The Beaufort Sea area appears to be a very attractive oil province.

The first OCS lease sale (Sale BF) in 1979 yielded $584 million for leases.   Three test wells were drilled in 1982.   All were dry, but they yielded information that oil was present and that there was adequate reservoir rock.   A well at Tern Island found oil in noncommercial amounts. The second lease sale in 1982 (Sale 71) brought more than $2 billion.   In 1983, a second well at Tern Island also found oil, but a well believed to be promising, Sohio No. 1 Mukluk, was dry.   Those in industry were disappointed by this result in the Beaufort province because Mukluk had been thought to have the potential of being an offshore equivalent of Prudhoe Bay.   The next year (1984), however, a discovery estimated at 300 million barrels of oil was made at Seal Island.   The potential of the province was thus established even though the discovery was too small to be commercial. Lease sale 87, held in 1984, yielded bids of $877 million.

Since 1984, 18 wells have been drilled to test various kinds of traps on the Arctic Platform.   The majority of the wells had shows of oil, although no accumulations large enough for commercial production were discovered. Additional lease sales in 1988 (97) and 1991 (124) yielded $118 and $17 million, respectively.   The low 1991 bid is more reflective of the depressed state of the oil industry and the operational and functional problems of working in the Beaufort Sea than it is an indication of the oil potential of the province.

The potential was emphasized in October 1992 when ARCO announced an oil discovery in its No. 1 Kuvlum well 97 km east of Prudhoe Bay in federal waters of the Beaufort Sea.   The exact size of the discovery will be determined by subsequent confirmation and delineation wells, but at this writing, it is generally believed to be at least 1 billion barrels.   If so, the commercial productivity of the province has been established and the nation has a new major oil reserve.

ADEQUACY OF THE GEOLOGICAL DATA BASE AND ITS TREATMENT BY MMS

Although masked by the rigors of its climate and with the additions of sea ice and permafrost, the Arctic Coastal Plain and the continental shelf of the Beaufort and eastern Chukchi share fundamental geological characteristics with other continental margins generated by rifting and seafloor spreading. Similarities with the Atlantic continental margin of the United States are many and striking. The North Slope and the adjacent continental shelf, slope, and rise are correctly identified in MMS reports as being "Atlantic-type." Although climate has imposed an unusual and harsh surficial environment, the general geology of the Beaufort-Chukchi OCS is neither strange nor unusual.

Geologists familiar with other areas of similar tectonic setting, therefore, can evaluate with some confidence the extent and quality of geophysical and geological information available for the Arctic OCS. The density of seismic coverage, the number and types of wells drilled, and the volume of data collected are impressive.

The geological data base for the Beaufort and Chukchi OCS has several components: scientific research published in the general literature; studies undertaken for government decisions, such as those found in MMS's Environmental Impact Statements (EISs); industry-sponsored studies for feasibility and design; industry research for satisfaction of permit or regulatory needs; and proprietary exploration studies and other company confidential activities. The result is an extraordinarily rich literature, much of it published in the past 20 years. Four lists of references in geology illustrate the point. MMS's "Geologic Report for the Beaufort Sea" lists about 200 references (Craig et al., 1985); that for the Chukchi lists about 100 (MMS, 1987d). The Environmental Studies Database System geology search conducted by the committee contains about 375 citations (MMS, 1992d). The "ARCO Arctic Environmental Reports Collection Catalog and Index" cites about 270 studies, many geological (ARCO, 1989). The literature also contains consultants' reports scrutinized by both industry and regulators, agency reports (both state and federal), and peer-reviewed journal articles.

Company proprietary data are, of course, not available to the public, but most of them pertain to exploration rather than to environmental matters. MMS receives these data in confidence and uses them in its analyses. Despite the large data base, the EISs for Beaufort and Chukchi sales contain surprisingly little geological information, which might create the incorrect impression that relatively little is known. Geological topics are covered in the EISs largely by citations to references, which must be consulted to appreciate the extent of geological knowledge of the area. The content of the Beaufort and Chukchi reports verifies the existence of a data base adequate to provide a basis for leasing decisions. The reports appear to be objective and in adequate detail. Moreover, the information base has been upgraded and experience has added information since the reports were written (in 1985 and 1987). The committee found no major contradictions between MMS's synthesis and the compilation of arctic geology entitled "The Arctic Ocean Region," a part of the Decade of North American Geology (DNAG) series of the Geological Society of America, published in 1990 (Grantz et al., 1990).

The ARCO Kuvlum No. 1 discovery in the Beaufort Sea can be interpreted as verification of MMS's resource estimates. The discovery, which is in OCS Sale 87, is currently estimated at 1 billion barrels of recoverable oil, a figure consistent with MMS's preliminary estimate. Indicative of its very conservative approach, MMS will not significantly revise its estimates for the Beaufort despite this discovery, because only one well has been drilled to date. MMS chooses to consider this a substantiation of its methodology rather than proof of a larger-than-projected resource. But if Kuvlum proves out, other known, but currently uneconomic, fields in the vicinity should become viable. Amerada Hess, for example, has announced that it plans to drill a well near its 1 Northstar discovery, which has been capped since it was drilled in 1985 (Davis and Pollock, 1992).

The only apparently pertinent differences among MMS's geological reports, papers in the DNAG volume (Haimila et al., 1990), and the derivative sections of EISs concern the importance of gas hydrates, which, if present in sufficient quantities and concentrations, can be a source of natural gas (methane). But in the Alaskan OCS, their presence could create engineering problems, especially regarding foundations for offshore structures. The current state of knowledge is illustrated by two quotations. "Gas hydrates beneath the Beaufort continental slope are not likely to be a significant problem in the foreseeable future since they occur at water depths in excess of 300 m, well beyond the operating limit for petroleum development" (Craig et al., 1985). "Sediments of the Arctic Ocean region may trap enormous quantities of natural gas in and beneath gas hydrates. These are found (1) offshore, in sediment of the outer continental margin, and (2) onshore, in and below areas of thick permafrost" (Haimila et al., 1990). Such statements might be compatible, but not enough information is given to verify them.

The committee concluded that the geological data base is adequate for the purpose of lease sales, but that the lack of adequate geological summaries in the EISs makes it difficult for other disciplines, such as oceanography and biology, to use the information in attempting to assess the potential impacts of OCS exploration and production.

RESOURCE ASSESSMENTS

Several analyses of the petroleum resources of the Alaska OCS region, and of the Chukchi Sea and Beaufort Shelf basins in particular, reveal a potential for future supplies. Although individual studies show somewhat different numbers for the magnitude of the resources, collectively they support the expectation that the region will continue to be a focus of interest for hydrocarbon exploration and development. The amount of interest and activity cannot be predicted with certainty; it will depend on economic factors, the relative attractiveness of the prospects in a global context, geopolitics, and supply-and-demand characteristics. However, the possible effects of exploration and development on the environment and the populace must be addressed.

The attractiveness of the Alaska OCS region for oil and gas production is based not only on estimates of the total resources, but even more on the probable sizes of future discoveries. This concept is well stated in Fisher et al. (1992):

> Alaska has the greatest potential for the discovery of major new oil fields in the United States. Fields in Alaska that are considered "marginal" in an economic sense are thought to contain immense reserves in excess of those discovered in any onshore field in the lower 48 states during the past few decades. Remaining unexplored or underexplored areas in the Alaskan North

Slope, both onshore and offshore, offer the best opportunities for oil discoveries in the giant and supergiant categories.

QUANTITATIVE ESTIMATES

A relatively small number of assessments of the oil and gas potential in the Chukchi Sea and Beaufort shelf basins were made available to the committee. The benchmark estimates were those prepared by MMS and included in the 1989 report of the Department of the Interior, "Estimates of Undiscovered Conventional Oil and Gas Resources in the United States—A Part of the Nation's Energy Endowment" (USGS/MMS, 1989). MMS's estimates come from continuing studies by the service that use a sophisticated computer-based probabilistic method called PRESTO (Probabilistic Resource Estimates—Offshore). MMS's estimates are presented in several forms: risked and unrisked, technically recoverable, and economically recoverable; and as a range of values reflecting three confidence levels (95%, arithmetic mean, and 5%). The 1989 estimates of risked recoverable oil and gas are listed in Table 3-2.

Another quantitative estimate of the total oil potential of the Alaska offshore region is contained in a report by Fisher et al. (1992), in which the mean potential is described as 2 billion barrels, assuming advanced technology and a price of $20/barrel; 4 billion barrels, assuming existing technology and a price of $27/barrel; and 8 billion barrels, assuming advanced technology and a price of $27/barrel. One must therefore assume an estimate of an implied unconstrained potential in excess of 8 billion barrels. This number is essentially twice that of the 1989 MMS study, presumably incorporating more recent evaluations by the industry.

The committee also had unpublished presentations made to it by representatives of ARCO, Chevron, and Mobil. Quantitative estimates were provided in the form of a bar chart, from which the committee derived Table 3-3.

These numbers suggest that at least some companies consider the mean resource potential to be considerably greater than those estimates contained in the MMS or the DOE studies cited by Fisher et al. (1992). The ranges

Table 3-2   1989 Estimates of Recoverable Oil and Gas

	Crude Oil, Bbbl			Natural Gas, TCF		
	95%	Mean	5%	95%	Mean	5%
Beaufort Shelf	0.49	1.27	3.74	2.14	8.26	12.81
Chukchi Sea[a]	0.00	2.22	7.19	0.00	6.33	16.87

[a] The approximate 50% discounting of the values for the Chukchi Sea is based on the assigned 0.5% risk that hydrocarbons will not be present in the region. (As oil and gas are discovered in the region, this risk would be removed.) The totals reported for the two basins would be approximately 3.5 Bbbl (billion barrels) and 14.6 TCF (trillion cubic feet) (mean values). Based on the 1989 report about one-third of the undiscovered recoverable oil resources of the United States were estimated to be located in Alaska; 20% of which would exist in the offshore basins. Source: USGS/MMS, 1989.

Table 3-3   Quantitative Estimate of Total Oil Potential for Beaufort and Chukchi OCS

	Crude Oil, Bbbl		Natural Gas, TCF	
	Mean	5% Probability	Mean	5% Probability
Beaufort Shelf	11	30+	15	65+
Chukchi Sea[a]	4.3	21.7	15	73

[a]The approximate 50% discounting of the values for the Chukchi Sea is based on the assigned 0.5% risk that hydrocarbons will not be present in the region. (As oil and gas are discovered in the region, this risk would be removed.) The totals reported for the two basins would be approximately 3.5 Bbbl (billion barrels) and 14.6 TCF (trillion cubic feet) (mean values). Based on the USGS/MMS 1989 report about one-third of the undiscovered recoverable oil resources of the United States were estimated to be located in Alaska; 20% of which would exist in the offshore basins.
Source: ARCO, Chevron, and Mobil, unpublished data, 1992.

of estimates are also shown to be greater by the much larger 5% values both for oil and gas, pointing out the large uncertainty in all current estimates. It is also noteworthy that in contrast to the MMS estimates, the potential of the Beaufort shelf is rated by some as being greater potential than that of the Chukchi Sea.

The company presentations to the committee provided important insight to the petroleum industry's perspective of the region. The finding of relatively rich, thermally mature rock (a sediment rich in carbonaceous matter is good source rock only after it is heated to convert the organic materials into fluid hydrocarbons; the rock is then called "thermally mature") in the Chukchi Sea and of pooled oil in subeconomic quantities plus the descriptions of the typical exploration plays in both regions all tend to verify the potential of the Beaufort shelf and Chukchi Sea regions. The economic difficulty in developing small accumulations of oil also was emphasized.

The diversity of the several estimates is not unusual considering the relatively small amount of exploration in the region. As discussed below, the MMS estimates were prepared with a conservative bias imposed by methodology, and they are the only fully documented and published results available. They are science based, they use properly documented methodology, and they are derived from a comprehensive data base. The companies' methods were not explained and could have introduced differences in addition to those that result from differing geological opinion.

ADEQUACY AND RELIABILITY OF ESTIMATES

In 1991, the National Research Council (NRC) Committee on Undiscovered Oil and Gas Resources produced an evaluation of the Department of the Interior's 1989 assessment procedures used in assessing the nations resources of oil and gas (NRC, 1991b). MMS's activities in the Alaska OCS were reviewed in that document, and the report can be used as a starting point for consideration of the status of information in Alaska. It must be noted that the focus of the 1991 report was not on the mainstream activity of tract evaluation for lease sale purposes, but rather on the "exploration- play"-level assessments that constituted the national assessment. Because the same basic methods and programs were used by MMS for tract-evaluation and exploration-play-level assessments (modified for the larger purpose), many of the observations in the 1991 report are relevant to the task of this committee. The next sections recapitulate some of the 1991 report's observations.

Data Base and Regional Studies

The Anchorage, Alaska, office of MMS has an excellent data base consisting of modern, high- quality seismic lines and structure maps for most of the area of current interest; data from exploration wells; and regional geological studies. Mapping is continuous, with revisions done by a competent and well-motivated staff. Individual prospects are identified and probable reservoir characteristics are compiled in an orderly fashion. One possible deficiency, common to several MMS offices, was a dearth of geochemical studies related to source rock abundance, organic carbon types, and thermal maturity.

Methods

The NRC Committee on Undiscovered Oil and Gas Resources found fault with several aspects of the way in which data and geological judgment were incorporated into the PRESTO model that is used to estimate resources. The PRESTO software is specifically designed to use prospect data to perform a variety of assessment tasks. Because there is considerable uncertainty associated with most of the data used as input, geologists must formulate probability distributions of reservoir variables and subjectively estimate several levels of risk. Concerns of the committee included the following:

• There was an apparent inconsistency in the methods used by different geologists and engineers to assign maximum and minimum values to the distributions. Although forms had been developed to assist in the compilation of data, there did not appear to be enough training in probability methods to ensure adequate results.
• The assignment of risk appeared to be poorly understood and unevenly applied. The multiplicity of risk levels (including economic thresholds) was considered overly complex, possibly leading to "double discounting" in

some cases. Although guidelines for assigning risk had been developed, the assignments appeared arbitrary in some cases and mechanical in others. No studies of the independence versus the dependence of risk variables had been done. Because risk assignment has such a profound effect on the final resource estimates, additional training was recommended.

• Several input variables were assigned inappropriate, arbitrary limits. These included limiting the allowable values of prospect areas and vertical fill-up; rejecting prospects from consideration that had less than 30 m of net play, that were at depths of less than 91 m, or that were smaller than one-half a lease block. Prospects that were judged to be subeconomic (if assigned appropriate risk) were not modeled in the conditional case.

• Some stratigraphic and conceptual plays were not considered adequately. The structural plays dominated the assessment, based on the mappable prospects. It appeared that "unseen" prospects, too small to be included in the existing seismic grid, were largely ignored.

• There was too heavy a reliance on the use of mean values instead of on descriptions of the full range of estimates that reflect the associated uncertainty in the final reporting of estimates. There also was speculation that the expression of "conditional estimates" could be confusing or misleading to the users of estimates.

The Committee on Undiscovered Oil and Gas Resources concluded that the likely effect of the deficiencies in the application of methods was an understatement of the resources, although the committee was unable to verify this conclusion. Industry presentations to the present committee tended to verify that MMS's estimates are conservative; this committee did not attempt a detailed comparison. (It may be noted that many of the previously noted shortcomings have been addressed by MMS.)

The present committee's charge has a very different focus from that of the committee that reported in 1991. Our focus relates to the adequacy of resource-potential information to support studies of environmental issues and concerns associated with offshore leasing. Specifically, this committee was asked to focus on the adequacy of resource information made available in EISs and SIDs (secretarial issue documents) associated with lease sales. The committee recognizes that the estimates prepared by MMS and provided in EISs are an adequate, albeit conservative, estimate of the quantities of oil likely to exist within a given lease area. The EISs, however, do not contain the detailed information that was used to generate these "bottom line" estimates, such as the number, magnitude, distribution, or characteristics of the individual areas assessed. There is no information regarding risk factors and economic criteria used in the lease-decision process. Because the gas resources likely to be present in the area were deemed uneconomic, no measure of this component is included in the EISs. Presumably, the proprietary nature of some of the input data used in assessment precludes complete disclosure of some elements of the output estimates. This is unfortunate because the very comprehensive oil and gas leasing-decision process used by MMS generates a lot of data and information that could be invaluable to those attempting to quantify the risks associated with environmental impacts.

The practice of resource assessment has matured with the development of programs like PRESTO and others that have led to probability-based estimates now being the "norm" rather than the exception. In most modern estimating procedures, not only the magnitude, but also the marginal probabilities or risks of occurrence associated with them are quantified. PRESTO effectively considers interrelated risks and input data in a simulation process that reflects the needs of industry for exploration in developing an estimate. If environmental scientists are to make comparable analyses, it will be necessary to continue or expand the simulation process to include quantification of both the likelihood and the magnitude of a variety of risks. They then could use the additional data from the assessment process in their analyses. Several of the data elements that could be required by environmental scientists include the following:

• The degree of concentration or dispersion of the resources. Are they likely to occur in numerous small-to-intermediate size fields or in a relatively few large deposits?

• The anticipated size of a typical deposit, along with some reservoir characteristics, such as field area, depth, temperature, and pressure.

• Product mix. Is gas likely to be associated or produced with any oil encountered?

• Risk information. What is the likelihood that gas and oil will be encountered in any given exploratory well?

• Formation fluids. Is there information on the nature of formation fluids that could be encountered?

• The type of drilling or production facility likely to be used.

Most of the items listed above are considered, used, or generated in the assessment process. If a mechanism were devised to provide greater disclosure of resource assessments, it might assist materially in the development of better models for environmental studies.

ENGINEERING GEOLOGY

Several types of geological engineering studies will be required as sites are selected for exploratory drilling. Even more extensive studies will be necessary if enough petroleum are discovered to warrant development. These studies are required by leasing agreements, follow accepted engineering standards, and are mentioned briefly in MMS's final EISs. The selection of the exact sites for exploration and production facilities will, of course, depend on the petroleum companies' estimates of the distribution of the resources and on the development plans for producing the fields. Two general types of studies will be required: foundation studies beneath and in the vicinity of the proposed exploration or production facility and studies along possible routes for subsea pipelines to bring the oil and gas to onshore transportation facilities. Because these studies are not only required but also are essential to protect equipment and personnel, industry will pursue them thoroughly.

Investigations in and around a drill site or production facility concern foundation stability, including the likelihood of slumping, the presence of subsea permafrost and gas hydrates, and physical properties of the formations. Such studies are required by leasing agreements and are part of current engineering practice; they are mentioned briefly in MMS's final EISs from Beaufort Sale 124 and Chukchi Sale 126. Slumping is a possible hazard in the Chukchi and Beaufort regions, but foundation difficulties can be overcome by proper design of the drilling platform or island.

Detailed studies also will be required for the route of any proposed pipeline. For much of the route, the pipeline will need to be placed below the seafloor and below the reach of ice scouring, including deep ice keels. The extent of permafrost beneath a veneer of soft sediment and the local presence of consolidated sediment and rock will require appraisal. If there is a choice of routes, studies of the geology could dictate the actual path.

Where pipelines come ashore to join shore installations or other pipelines, geological conditions will require evaluation. For example, their effect on shorelines, beaches, bars, lagoons, and areas of focused ice action will require study; the route and design of the pipeline will depend on these studies being considered.

In 1983, the NRC Marine Board reviewed what was known about the topics covered by such studies and concluded that there was "no area where lack of seafloor technology would preclude safe operations offshore or delay offshore development" (NRC, 1983b). The report noted also that the arctic seafloor poses special engineering challenges and that there are subject areas in which more information is needed. Seafloor conditions that will greatly influence the design and economics of offshore activities include thaw subsidence (associated with the presence of subsea permafrost), sedimentary erosion and transport, ice gouging of the seafloor, and the soil mechanics of silts. Details and an extensive reference list can be found in the 1983 report. This committee notes that there have been no recent major advances in many of the subjects.

Permafrost and Thaw Subsidence

Information about the strength of highly saline, warm subsea permafrost and about thaw subsidence is limited. This will undoubtedly dictate conservative design of offshore facilities. Subsea permafrost is a significant problem on large areas of the Beaufort shelf where its properties are extremely variable. The subsea permafrost appears to be thinner and limited to within 2 km of the coast in the Chukchi Sea; it apparently is absent in the Bering Sea except for the most rapidly retreating coastal segments of Norton Sound. Even in the Beaufort, detailed information on the distribution of subsea permafrost is limited. Recent studies in both the United States and the Canadian portions of the Beaufort have produced advances in electrical and high-resolution seismic methods for mapping subsea permafrost. These techniques will undoubtedly prove useful to future oil development.

Sedimentary Erosion and Transport

Because of the ice content of coastal sediments, rapid coastal erosion and movement of large amounts of sediments are common and highly variable. This is particularly true in the Beaufort Sea, where coastal erosion is a major problem. Quantitative information on these processes will be necessary for regions where causeways or pipeline crossings are anticipated. Sediment transport studies would contribute knowledge also in coastal-erosion problems and the infilling of ice gouges. (See U.S. Geological Survey Map I-1182-H, USGS, 1992b.)

Soil Mechanics of Silts

Silt-sized materials are common at many Alaskan OCS sites, and their properties are not understood as well as are those of sands and clays. Site-specific studies will need to characterize the specific properties of these materials. Another somewhat unusual characteristic of the seabed sediments in the Beaufort is that they are commonly overconsolidated; they are more compacted than would be expected. Because the mechanics of overconsolidation are not well understood, it is difficult to predict the distribution and the behavioral physical reactions of the material. Site-specific studies should pay particular attention to the response of these materials to cyclic loading and to the effects of freeze-and-thaw cycles on their strength and permeability.

In most cases, studies by MMS and industry have established an adequate framework for preliminary offshore designs of exploration and production facilities. As a result, with the few exceptions mentioned above, additional regional studies of the seafloor of the region are not warranted until specific drilling production sites are selected. At that time, however, careful, site-specific foundation and pipeline route investigations will be needed.

The geology of the Chukchi and Beaufort seas is reasonably well known and adequately discussed in MMS's final EISs. No additional regional studies of the seafloor in the vast Arctic region are warranted until drilling sites are selected. At that time, careful foundation and pipeline- route investigations will need to be done. This should be ensured by industry for design and safe execution and by government for environmental protection.

CONCLUSION AND RECOMMENDATION

The specific conclusion and recommendation for this chapter follows. For the general and overall conclusions of this report, see Chapter 8.

Conclusion: The committee concludes that MMS has developed resource estimates based on adequate data and modern techniques that give credible, but conservative, estimates of OCS oil and gas resources. The extensive resources, a significant portion of which remains in the nation, ensure that there will be continued interest in leasing, exploration, and development. The geological and resource-base information developed by MMS can provide a useful foundation for assessing the effects of exploration and development on the biological, physical, and human environments.

Recommendation: *Expand geological and resource base information in EISs. Additional information that would assist environmental scientists in quantifying risks should be released and included in EISs.*

Alternative: Call attention to the geological reports that address the potential size and distribution of oil and gas resources and make them readily available.

4 PHYSICAL ENVIRONMENT

INTRODUCTION

This chapter briefly reviews the meteorological, oceanic, and cryological (ice) characteristics of the environment in three (OCS) lease-sale areas in Alaska: the area in the Navarin Basin portion of the Bering Sea and the two in the Chukchi and Beaufort seas (see Figure 1-1). (Although this report focuses much more on the Chukchi and Beaufort lease areas than it does on the Navarin Basin, the Bering Sea is important because it interacts with the Chukchi Sea and, to a lesser extent, with the Beaufort Sea.) The adequacy and applicability of existing studies are discussed as they apply to environmental characterization from the point of view of offshore oil and gas development. Specific recommendations about investigations that would contribute useful information about each of regions are given in Chapter 7.

The high-latitude locations of the three lease-sale areas make them fundamentally different from most lower latitude regions. Their environment is quite inhospitable for temperate-zone people for much of the year because of extreme cold, strong winds, darkness, and sea ice. However, the environment is strongly seasonal. The Navarin Basin can be ice-covered in winter but is more temperate in summer, when all the sea ice usually melts. Some portions of the Chukchi and Beaufort seas are affected by sea ice all year, containing both seasonal first-year ice and thicker multiyear Arctic pack ice. The possibility of cross-boundary (or cross-border) impacts on Canadian and Siberian waters always exists. However, there is a stronger possibility of transport into U.S. waters from Siberian

(Bering Sea and Chukchi Sea) and Canadian (Beaufort Sea) coastal shelf waters because of the mean flow patterns.

All three lease areas are remote. Although there are many people and much equipment on the North Slope, offshore locations can be hundreds of miles from help, as are some onshore locations. Even if help is available close by on land or on an island, the harsh environment can prevent transportation to remote drilling or pumping sites or to a ruptured oil line or disabled tanker. Furthermore, the United States has only two open-ocean, polar-class, ice-breaking ships, the Coast Guard's *Polar Sea* and *Polar Star*. It is possible, at any given time, for one or both of these ships to be deployed in the Antarctic or in the Atlantic sector of the Arctic and to be unavailable for help in the Alaskan OCS. A possible tempering consideration is that the United States and Russia have an agreement for mutual aid in case of an oil spill; the Soviet Union offered assistance after the *Exxon Valdez* spill in Prince William Sound.

The background that follows could give the impression that there is an abundance of data about the physical environment of the Bering, Chukchi, and Beaufort lease-sale sites. In reality, the region is enormous, and the data are probably barely adequate to correctly predict which way the air, ice, or ocean will move at any one time or place, say, in the event of an oil spill. The background on ice is found in Chapter 7. For recent reviews of these regions, see Carmack (1990) and Niebauer and Schell (1993).

We note that although the arctic environment is hostile, Alaska Natives have lived there for at least hundreds of years and continue to do so. A great deal of their food and materials comes from ice-infested ocean waters. Their knowledge of the physical and biological environment is critical to their survival. Under certain circumstances and in specific places, a good hunter might better predict a spill trajectory than a general circulation model (GCM). Although most Alaska Natives' knowledge is not written down and is therefore not accessible in a data base, traditional knowledge is an important counterpart to Western science and should be taken into account seriously.

ENVIRONMENTAL STUDIES

Physical oceanographic studies (including field studies and modeling in oceanography and meteorology) provide a basis for predicting oil spill transport in OCS regions. The starting point is typically output from an ocean circulation model, forced by observations or model data on meteorology that are used to compute possible oil spill trajectories from selected points. In conjunction with the probability of an oil spill for selected launch sites, an oil spill risk analysis (OSRA) model is used by the Branch of Environmental Operations and Analysis (BEOA) of MMS to estimate the subsequent probability of the oil's arriving in sensitive areas or on the shoreline within a given time. It is the compilation and distillation of these probabilities that find their way into an Environmental Impact Statement (EIS). OSRA calculates only the movement of the center of mass of each hypothetical spill. Separate estimates are made of the surface area actually covered by oil from a spill of a particular size and of the much larger area over which the oil would be discontinuously spread. The details of oil transport models are provided in the EISs and references in them.

ARCTIC OCEAN

Arctic Ocean water and ice circulation are largely contained and restricted from contact with the temperate oceans by the continents and islands that surround it. The center of the Arctic Ocean is permanently ice-covered; the periphery is seasonally ice-free. It is the seasonally ice-covered periphery that includes the lease-sale sites in the Chukchi and Beaufort seas. Should a spill occur, the ice, as well as the currents in the region of these sites, will carry spilled oil.

The Chukchi and Beaufort seas are off the north coast of Alaska in the Arctic Ocean proper; the Bering Sea is subarctic, located between Siberia and Alaska south of the Bering Strait and north of the Aleutian Island chain. The Bering Strait is the only direct connection to the Arctic Ocean from the Pacific Ocean; it is about 50 m deep and 85 km wide. The net flow through the strait is northward and accounts for about 15% of the inflow to the Arctic Ocean (Aagaard and Greisman, 1975).

The Chukchi Sea is underlain by a broad shelf, nearly 900 km from the Bering Strait north to the shelf break. In comparison, the Beaufort Sea has a relatively narrow shelf. Proceeding east from Point Barrow, considered the boundary between the Chukchi and Beaufort seas, the continental shelf narrows to about 70 km. The shelf widens again farther east near Mackenzie Bay to about 160 km and extends eastward into the Amundsen Gulf.

The marine and submarine topography of this Arctic Ocean shelf region includes Wrangel Island, at the approximate western boundary between the

Chukchi and East Siberian seas, and Herald and Hanna shoals in the Chukchi Sea shelf north of Bering Strait. Submarine canyons include the Herald and Hanna canyons in the Chukchi Sea, and the Barrow Canyon at the boundary between the Chukchi and Beaufort seas. There are also numerous barrier islands along the north and Chukchi coasts of Alaska. Approximately the northeast half of the Bering Sea (which contains the Navarin Basin) overlies the widest continental shelf in the world outside the Arctic; the southwest half overlies an abyssal plain with depths of about 4 km. The eastern Bering Sea shelf is about 500 km wide, but it is shallow (approximately 170 m at the shelf break). This shelf borders most of Alaska as well as the coast of Russia north of Cape Navarin. Southwest of Cape Navarin, the shelf narrows by an order of magnitude. The continental slope is indented by several undersea canyons. King, St. Lawrence, St. Matthew, Nunivak, and the Pribilof islands are in the eastern Bering Sea shelf. The climate and weather of this high-latitude region are strongly related to the presence and fluctuations of sea ice (Overland, 1981), which are related to the large seasonal variation in insolation. The Arctic is surrounded by the large weather systems of the northern hemisphere. In summer, when the North Pacific high, the Asian continental low, and the North Atlantic high expand to the north, the weather in the Arctic is relatively moderate. In winter, the patterns change drastically so that the Asian continent is dominated by high pressure, and the North Pacific and North Atlantic are dominated by the Aleutian and Icelandic low-pressure systems, respectively.

Because of the weather pattern change, including the intensification of the Aleutian low, winter is harsh in this part of the Arctic. The mean position of the Aleutian low is actually a statistical artifact; the low is the average position through which most of the migrating cyclones pass. Three to five storms each month pass along the Aleutians into the Gulf of Alaska or into the Bering Sea-Bristol Bay region. In comparison, fewer than two storms a month pass across the northern Bering (Overland, 1981). Relatively few storms occur north of Bering Strait. Brower et al. (1977a,b) showed an average of fewer than one each year for any given month for the region from Wrangel Island eastward to east of Mackenzie Bay. Although infrequent, the storms that do occur in this region can result in storm surges and severe wave action along the north coast of Alaska, especially during the late summer and early fall. Intense storm surges occur about once each year (Brower et al., 1977b). To the north of Alaska, the interaction of different weather systems surrounding the Arctic generates a poorly defined relative high-pressure system with clockwise circulation over the Beaufort Gyre, which drives the clockwise flow of the near-surface ocean and ice in the Beaufort Gyre. The major rivers that enter the region of interest include the Yukon River in the northern Bering Sea/Norton Sound, the Colville River in the U.S. Beaufort Sea, and the Mackenzie River in the Canadian Beaufort Sea. There are a number of smaller rivers such as the Kobuk and Noatak that empty into the Chukchi and Beaufort seas.

As pointed out by Carmack (1990), there has not yet been a synthesis of arctic shelf waters, although he listed many individual studies. Carmack pointed out that annually the shelf waters go through a much larger variation in salinity (~2-4 parts per thousand (ppt)) than do the surface waters (~0.5 ppt) of the Arctic. The shelf waters are less saline in summer due to river inflow and ice melt but are more saline in winter due to river freeze-up and brine rejection from ice formation. In addition, in winter, upwelling may cause higher salinities on the shelves, reaching 34.5 ppt or greater (Aagaard et al., 1981; Melling and Lewis, 1982). During the summer, the shelves show net offshore flow at the surface similar to an estuary while in the winter, there is net inflow of more dense water at depth (Macdonald et al., 1989; Carmack et al., 1989; Carmack, 1990).

Thus, the Arctic Ocean surface layers are freshened by river and glacial input, which are strongly seasonal in nature (Carmack, 1990). The two major freshwater inputs to the regions of interest here are the Mackenzie River (340 km^3/yr or 0.01 Sverdrup (Sv; 1 Sv = 10^6 m^3/s), which flows into the Beaufort Sea, and the Yukon River (214 km^3/yr or 0.0068 Sv), which flows into the eastern Bering Sea. The seasonal variability for the Mackenzie River is about 5-fold and it is about 12-fold for the Yukon River. In comparison, about 1,500-2,000 km3/year (~0.05 Sv) of fresh water enters the Arctic Ocean as a component of the inflow through the Bering Strait, which includes some of the Yukon River.

SOUTHEASTERN BERING SEA

The physical environment of the Bering Sea shelf is characterized by strong variability in the air, ice, and ocean climate regimes. This includes periods of days to years, including a pronounced interannual variability (Niebauer, 1988). Daylight is nearly nonexistent in winter and nearly continuous in summer. The wind stress over the sea also varies by an order of magnitude from summer to winter. The shelf region is ice-covered in winter, although the entire Bering Sea is ice- free in summer.

Tidal currents dominate the entire OCS; they contribute 60% of the horizontal kinetic energy in the outer shelf (from water depths of approximately 100 m to the shelf break) to 90% along the coast (Kinder and Schumacher, 1981a).

Because the southeastern Bering Sea shelf is so broad, the various sources of energy, such as tides, winds, and freshwater input, are applied over a large area. Over the southeastern shelf, the mean current flow is low, 0.01-0.05 m/s, and moves toward the northwest. Seaward of the shelf break, the Bering Slope current flows at speeds of approximately 0.1 m/s with numerous eddies (Kinder and Schumacher, 1981b). Because of the slow mean flow, the hydrographic structure on the shelf tends to be locally formed by the input from insolation, cooling, melting ice, freezing, and river runoff, as well as lateral exchange with bordering oceanic water masses (Kinder and Schumacher, 1981b; Coachman, 1986).

The Bering Sea has four distinct hydrographic domains, each with associated current patterns, defined by water depths and boundary fronts that are generally parallel to the isobaths (Kinder and Schumacher, 1981a,b; Coachman, 1986). They are the coastal, middle shelf, outer shelf, and oceanic domains.

NORTHERN BERING SEA AND BERING STRAIT

The Bering Strait is narrow (85 km) and shallow (50 m). Flow averages about 0.25 m/s and about 1 Sv in summer, and 1-1.5 Sv in winter (Coachman and Aagaard, 1988). The northward flow results because the sea surface tilts downward toward the north. There are strong currents and current shears across the strait. The strongest currents—in the upper layers of the east side—can be more than 2 m/s. The flow in the western side can be 0.5-0.6 m/s with little vertical shear. However, reversals in flow of at least a week's duration, especially in winter, are related to atmospheric pressure gradients over this region (Bloom, 1964; Coachman and Aagaard, 1981). Winds tend to be channeled north or south through the strait.

In the northern Bering Sea (approximately north of 62° N including Norton Sound), there are three identifiable water masses and related fronts. The water masses are related but not identical to the water masses in the southeastern Bering Sea (Coachman, 1986; Hansell et al., 1989). Fronts between these water masses are identified (Paquette and Bourke,

1974; Wiseman and Rouse, 1980; Muench, 1990) as associated with the boundaries between the water that flows from the Bering Sea through the Bering Strait into the Chukchi and Beaufort seas and the Arctic Ocean, driven by the sea-level difference between the Pacific and Arctic oceans. These waters make their way eastward along the Alaskan arctic coast as outlined below in the individual regions.

CHUKCHI SEA

The Chukchi Sea widens north of the Bering Strait with Kotzebue Sound immediately to the northeast of the strait. The shelf is relatively shallow (20-60 m). Herald Shoal, about 200 km due north of the strait, is 20-30 m deep. Many of the capes and headlands in the region on the Alaskan side of the Bering Sea are high mountains, which tend to cause "corner-effect" accelerations in winds along the coast, analogous to the high winds caused in cities by tall buildings (Kozo, 1984).

The three water masses that flow northward through Bering Strait cross the Chukchi. Paquette and Bourke (1981) showed that flow trajectories crossing the Chukchi are steered through the troughs and canyons. These flows result in melting that causes embayments in the sea-ice cover (see Chapter 7). North of the strait, the Anadyr and Bering water tend to combine and flow northward and slightly eastward at 0.15-0.20 m/s, bifurcating in the region between the Point Hope-Cape Lisburne headlands and Herald Shoal. Some of this water goes through the Herald Canyon into the Arctic Ocean to the west of Herald Shoal, and some turns eastward as Bering Sea water that flows eastward along the outer shelf of the Beaufort Sea (Coachman et al., 1975; Aagaard, 1984). In the eastern Chukchi Sea, Alaska coastal water also flows along the Alaska coast, gaining fresher, cooler water from the large rivers that empty into Kotzebue Sound. The Alaska coastal water flows at speeds of 0.25-0.30 m/s past Point Hope, Cape Lisburne, and Icy Cape toward the Beaufort Sea beyond. Water also "drains" from the Chukchi Sea through the Barrow Canyon off Point Barrow into the Arctic Ocean (Aagaard, 1984).

The maximum amplitude of tides in the Chukchi and Beaufort seas is only 5-20 cm, which is less than 1% of the total variancein local sea level; tidal speeds are about 5 cm/s, which is 1-2% of the total variance (Kowalik, 1984; Aagaard et al., 1989). In the coastal lagoons, tidal variation can be up to 2 m (Kowalik and Matthews, 1982).

BEAUFORT SEA

There are several subscale (kilometers to tens of kilometers) wind phenomena that are important to the shelf circulation of the Beaufort Sea (Kozo, 1984). Monsoonlike winds occur during the summer, caused by a semipermanent arctic atmospheric front that results from horizontal thermal contrasts along the coast. Heating of the land causes a deficit in atmospheric pressure, leading to onshore winds, which are turned to the right (toward the west) along the coast as a result of Coriolis acceleration. Related to the monsoon is a breeze (Kozo, 1984). Along the Beaufort Sea coast, sea breezes are characterized by large diurnal sea-to-land temperature contrasts, clockwise rotation of surface winds, and surface winds that oppose offshore gradient winds. At least 25% of the time the surface-wind direction is dominated by the sea breeze. One effect of these winds is the maintenance of coastal currents (0-20 km wide) toward the west, causing lagoon flushing. Sea breezes cause a general masking (about 25% of the time, as mentioned above) of synoptic wind conditions in this first 20 km from the coast. During summer, sea breezes along this coast are not followed by land breezes because the land stays warmer than the water (the summer sun does not set at this latitude).

Orography

The Brooks Range of mountains, which has a mean height of 1.5 km, is some 240 km inland of and parallel to the general trend of the arctic coast of Alaska. This major mountain range affects winds over much of the Beaufort Sea coast (Kozo, 1984). For example, near Barter Island, orographic modification of the winds can cause wind speeds 50% greater than that of the geostrophic wind (the wind resulting from a balance between horizontal pressure gradients and the Earth's rotation) because of the corner effect of the mountains, which are close to the sea. This effect can be felt as much as 350 km away, and it can influence the circulation of ice and water in the Beaufort Sea.

In addition, mountain barrier baroclinicity (a condition in which surfaces of equal pressure are inclined to surfaces of equal density) occurs when stable air moves toward and up a mountain without heating from below. This causes the isobaric (equal pressure) and isothermal (equal temperature) surfaces to tilt away from the mountain range, resulting in geostrophic winds parallel to the axis of the mountain range. This is mainly a winter phenomenon and is a major reason for wintertime winds from the west-southwest between Prudhoe Bay and Barter Island. This phenomenon results in a nearly 180-degree difference in wind direction when compared with the winds predominantly from the northeast at Barrow. Because mountain barrier baroclinicity depends on high surface albedo (the fraction of incident electromagnetic radiation reflected by a surface), it disappears in summer, when there is no snow. This effect has a horizontal extent of approximately 120 km and occurs about 25% of the time in the coastal zone from Prudhoe Bay to east of Barter Island.

Finally, wintertime atmospheric temperature inversions are common in the Arctic. They cause strong atmospheric surface stability that leads to diminished vertical turbulent exchange and hence to reduced wind stress on the surface to drive ice and ocean circulation.

Currents

Along the Beaufort Sea coast, in the event of a spill, the ice pack will trap and carry the oil in addition to the currents. There are also river plumes from the Colville River (and the Mackenzie River to the east in Canada), as well as the smaller rivers along the Beaufort coast, that affect the coastal and shelf flow of the the Beaufort Sea. The fresh water flowing out over the salt water causes buoyancy-driven circulation. In the winter, river flow often does not completely stop so there is freshwater flow out onto the open shelf under the land-fast ice. Variable weather patterns can cause these buoyancy flows to interact with winds, causing transient current jets along the coast, especially in summer when ice may not shield the ocean surface from wind stress. Shoreward of the 50-m isobath, local winds dominate the shelf flow of the ocean currents and drive them mostly toward the west under the generally prevailing easterly winds. However, there are periods, occasionally prolonged in some summers, when west winds cause easterly flow. Thus, the circulation at the coast is strongly wind-driven, but variable and highly seasonal; it is less energetic in winter because of the ice cover. The result is that significant coastal flow in this shallow inshore region appears to be primarily a summer phenomenon (Aagaard, 1984).

Kowalik and Matthews (1983) showed evidence that salt rejection from growing sea ice in a shallow coastal lagoon induces a two-layer water system.   The higher-salinity, higher-density, and, hence deeper water moves offshore at a rate of approximately 1-2 km/day (1-2 cm/s); the less-saline, less-dense upper layer moves onshore.   Drifter studies cited by Kowalik and Matthews (1983) showed shoreward movement for drifters released under the sea ice as far as 10 km seaward of the barrier islands.   Tidal and surge currents account for most of the variation in the currents, but these are superimposed on the mean brine-induced currents.

Offshore, seaward of 50 m over the OCS and continental slope, there is an organized band of flow toward the east, called the Beaufort Undercurrent (Aagaard, 1984), which is topographically steered but apparently not driven by local wind stress.   It is estimated that less than 25% of flow variability below 60 m is caused by wind (Aagaard et al., 1989).   The dynamics of the undercurrent are not yet fully understood, but it is characterized by a temperature maximum associated with eastward flow originating in the Bering Sea.   The Bering Sea water can be traced at least as far east as Barter Island.   This flow seems to be trapped along the OCS and slope between the 50-m (~40 km offshore) and the 2,500-m isobaths (~120 km offshore). Aagaard (1984) suggested that the Beaufort Undercurrent extends the full length of the Beaufort shelf and slope.   The mean currents are approximately 0.1 m/s toward the east.   Aagaard (1984) also reported frequent cross-shelf flow (mostly offshore flow of up to 5 cm/s for as long as 3 days), which links the nearshore region to the undercurrent.   The ocean driving of this circulation includes shelf waves and eddies (Aagaard et al., 1989).   Upwelling appears to be connected with eastward-traveling wavelike disturbances in the velocity records with vertical displacements of about 150m.

Continuing farther offshore of the Beaufort Undercurrent into the Arctic Ocean, the surface currents and ice flow are characterized by a mean westward movement of ice and water at the outer edge of the anticyclonic Arctic Ocean gyre.

SEA ICE

The presence of ice in the arctic OCS is arguably the most significant physical condition to be dealt with in developing OCS oil and gas resources (see Chapter 7).

Sea ice in the area shows great seasonal and interannual variability as well as some predictable mesoscale features (Burns et al., 1981). For example, the seasonal sea ice advance and retreat in the Bering Sea is the largest of any found in the Arctic or in subarctic regions, averaging about 1,700 km (Walsh and Johnson, 1979). Interannual variability in the position of the average ice edge in the Bering Sea is as great as 400 km (Niebauer, 1983). Farther north, the oceanic flow that fans out in the Chukchi Sea after going through Bering Strait causes significant mesoscale embayments in the Chukchi ice cover (Paquette and Bourke, 1981), which are predictable from summer to summer. Finally, winter conditions in the Chukchi and Beaufort seas produce fast ice along the coast that interacts with the offshore open Arctic Ocean current and wind-driven free-floating sea ice to cause an extensive, somewhat predictable, system of flaw leads and polynyas off the Chukchi and Beaufort coasts eastward to the Canadian Archipelago.

Maximum sea-ice cover occurs in March or early April, lagging minimum insolation in late December by 3 months because of the heat capacity of the ocean and the cold atmosphere. At this time, essentially all of the Arctic is ice-covered, as is about one-third to one-half of the Bering Sea. The mean ice edge in the Bering Sea is about 900 km south of Bering Strait, and it varies from 700 km to about 1,100 km south of Bering Strait, a range of 400 km, or more than 40% of the seasonal ice cycle (Niebauer, 1983).

Maximum retreat of the sea ice occurs in September, again lagging maximum insolation by about 3 months. In the mean, the ice edge retreats about 1,600 km between March and September, moving into the Chukchi Sea (the Bering Sea becomes ice-free). By September, in normal years, the ice pulls away from the Arctic coasts of Canada, Alaska, and Siberia, except for an approximately 300-km-long section of the Siberian coast; leaving a nearly continuous, relatively ice-free corridor around the permanent ice pack. In most years, this corridor varies in width from about 300 km at the western end of the East Siberian Sea to less than 50 km off Prudhoe Bay. The deviations around these locations are large. In warm years, the ice is 600 km off the coast of the East Siberian Sea and 300 km off the Beaufort Sea coast at Prudhoe Bay. In cold years, the ice sometimes does not pull away from the coast, although there is invariably open water in some of the bays along the north coast of Russia in the Chukchi and East Siberian seas. However, individual storms can cause large changes in ice

cover in a short time. For example, off Barrow, northwesterly and northerly winds can cause compaction of drifting ice, closure of open coastal
water, and extensive runup and pileup of ice onto the coast (Weeks and
Weller, 1984). Such events trapped and destroyed many whaling ships in
the northeastern Chukchi Sea during the late nineteenth century.

In September, when there is a minimum of ice, the distribution of ice and
open water reflects surface ocean flow patterns into the Chukchi Sea. Open
water generally reaches through the strait into the Chukchi and Beaufort
seas, paralleling the coast of Alaska north to Point Barrow and then
extending eastward all the way to the Canadian Archipelago. This is a
region of extensive nearshore leads and recurrent polynyas, which form as
a result of the interaction of the fast ice and land with drifting ice driven by
ocean currents and winds. To the west, the ice sometimes does not pull
away from the Chukchi coast south of Wrangel Island. During years of
unusually heavy ice cover the ice margin in September can be as far south
as about 66° N on the Siberian side but only 70° N on the Alaskan side.
This is the result of warm currents that flow through Bering Strait, hugging
the Alaskan coast to Point Barrow and then curving off to the east (Paquette
and Bourke, 1981). Paquette and Bourke (1981) depicted current flow
crossing the Chukchi as being steered by the bottom troughs and canyons.
This flow of warmer water from Bering Strait results in the melting of sea
ice and causes recurring embayments in the ice (Paquette and Bourke,
1981). In spring, along the northwestern coast of Alaska from Point Hope
to Point Barrow, there is a region of leads and polynyas offshore of the
land-fast ice (Stringer et al., 1982).

Most of the ice in these regions is open pack ice driven by wind and
ocean currents. The pack ice is primarily first-year ice, except in the Beaufort Sea, which contains some multiyear ice from the Arctic. Limited fast
ice is present in the eastern Bering Sea, found mainly in protected bays or
along shores facing the prevailing winter winds. In the Beaufort Sea, fast
ice is more extensive because of protective barrier islands and because the
grounded pileups of sea ice on the shelf act like small barrier islands.
However, the fast ice in the Beaufort Sea is seasonal, usually lasting from
November through June.

Ice drift rates are highly variable in the Bering Sea, frequently with rates
of 17-22 km/day. Rates as high as 32 km/day have been reported (Shapiro
and Burns, 1975; Muench and Ahlnas, 1976; Weeks and Weller, 1984). In
Bering Strait, ice movements of 50 km/day have been observed, although
there are reversals. In the Chukchi Sea, drift rates are considerably lower,

0.4-4.8 km/day for the mean annual drift, but rates as high as 7.4 km/day have been observed (Weeks and Weller, 1984). In the Beaufort Sea, the ice speeds are 2-8 km/day. The offshore ice generally follows the east-to-west anticyclonic ocean circulation, essentially parallel to the coast. Undeformed ice thickness increases toward the north from about 0.5-1.0 m in the open Bering Sea and 1 m in Bristol Bay to 2 m off the arctic coast (Weeks and Weller 1984). The Bering Sea ice cover is almost entirely first-year ice; north of Bering Strait, the ice is 25-75% second-year ice or older. The older ice is thicker, with a mean of roughly 4 m. However, pressure ridges of heavily deformed ice have been observed with maximum depths (keels) of 50 m and heights (sails) of 13 m. When they become grounded, these deep keels can cause appreciable gouging of the shelf. In near-coastal zones where grounded ridges can form, sails can be as much as 20 m high. The pack ice over the arctic shelf is commonly highly deformed, with up to 10 ridges/km, because of shearing. Farther offshore, beyond the shelf, 2-3 ridges/km is more typical. There are fewer large ridges in the Bering Sea ice because of the less constrained nature of the ice drift (there are fewer immovable barriers the ice can work against to form ridges as it drifts toward the open sea).

PHYSICAL OCEANOGRAPHIC STUDIES

Given the state of knowledge as described previously, we have come to the following evaluations of the available data.

Models of Circulation and Oil Spill Trajectories

OSRA is the model used by BEOA to generate oil spill trajectories from selected hypothetical spill points, and estimate the number of "hits" on an environmental resource target or a shoreline segment, and the conditional probability of some effect on the resource within a selected time. It is the compilation and distillation of these probabilities that finally find their way into an EIS.

The input requirements of the model include data on air and ocean circulation in an area. These inputs typically are taken from ocean circulation and meteorology models. Trajectories for all OCS waters except those in the Alaska region are calculated by BEOA. In the Alaska region, contractors calculate trajectories and provide them to BEOA for model input. An important consideration in Alaskan modeling is the presence of ice. The trajectories, and hence OSRA's results, have been of varying quality over the past 17 years (NRC, 1990a [Physical Oceanography]).

Circulation Models

The National Research Council has reviewed the information available for gas and oil leasing in other OCS areas (NRC, 1989a [California and Florida]; NRC, 1991a [Georges Bank]) and has concluded that ocean spill trajectory estimates have relied too heavily on GCMs. The committee believes the same conclusion holds for the three lease areas involved here. With the relative lack of observations, initial reliance on model predictions is understandable, but as stated in the previous NRC reviews (NRC, 1990a, 1993a) trajectory predictions must be tied more closely to observations than has been the practice. The committee also notes that GCMs for the Alaska region have been used by several different contractors using different models, with little effort to synthesize the various results or to reconcile the predictions with the limited existing observational data.

The committee believes that little progress can be expected from refining existing GCMs for such vast areas. It would be more useful to develop limited-area GCMs for use at sites selected for exploration and development. These could be used to explore the possible effects to biologically or ecologically important areas in the vicinity of the production zone, to explain how mitigation measures might work, and to predict oil movement in the event of an accident. It is important that such a modeling effort be coordinated with observational efforts in the same area and take advantage of traditional knowledge.

The duration of the existing base of observations, in addition to being geographically sparse, is insufficient to distinguish mean circulation from fluctuations that result from annual or even seasonal processes. MMS has taken an important step in addressing these issues in other lease-sale areas by deploying meteorological buoys. MMS is to be commended as well for equipping some of these with acoustic Doppler current profilers, which record the current profile beneath the buoys as a function of time. Because harsh weather and ice preclude the deployment of similar tools in most of the lease-sale areas, the description of long-term variability is an important issue.

To maximize the ability to contain spills and protect sensitive resources from them, a model must be used quickly enough to provide actual predictions of trajectories. This is different from the requirement in the EIS process for statistical estimates of oil spill trajectories in that it must be run on a time scale commensurate with an actual spill. It must include environmental data and forecasts for currents, wind, and ice conditions as observed during the spill, although it might contain a statistical component that would allow for uncertainties and predict the most likely or possible paths. Several such models are available and have been useful (Giammona et al., 1992). Once possible development sites have been selected, one or more of these models should be selected for use in case of an accident, and the appropriate site-specific information required by the model (such as local mean currents, tidal currents, mesoscale eddy energy, and ice conditions) should be measured and made ready for model input.

Atmospheric Forcing of the Ocean Circulation

The atmosphere is important in determining the local effect of wind stress and heat transport and also, remotely, as conditions in distant areas are propagated through the coastal ocean by pressure gradients. In OCS areas off the continental United States and in the Bering Sea, MMS has established a network of meteorological buoys to monitor the lower atmosphere over long periods (10 years). There is no comparable set of observations for the Chukchi and Beaufort lease-sale areas, although the committee believes that sufficient information is available from standard numerical weather forecasting products and from fixed-station observations to describe atmospheric variability on scales larger than the mesoscale (100 km). The absence of information related to mesoscale variability in the lease-sale areas is not seen as a flaw in the existing EIS, but that information will be required once specific production areas are identified.

Studies should be designed to determine the spatial structure of the wind field on scales of a few kilometers in areas of production, with emphasis on the mechanics of the marine boundary layer and the interaction between the layer and coastal topography. Such observational studies should involve aircraft surveys and fixed-station observations. To take advantage of the long time series of weather on larger scales, there should also be studies that define the relationship between mesoscale variability and fluctuations in larger scale weather.

Availability and Suitability Of the Observational Base

Considerable efforts have been made by MMS and other federal agencies, by the state of Alaska, and by local organizations such as the North Slope Borough to acquire observations on ocean circulation in the areas considered here. Although these efforts have in many cases been quite successful, the number of observations available to describe the circulation in this vast area is much smaller than it is for other OCS areas bordering the continental United States. Important distinctions also exist between the three lease areas: The Bering Sea, in which the Navarin Basin is located, is the area for which the richest base of observations is available; less information is available for the Chukchi Sea, and the greatest uncertainties concern the U.S. portion of the Beaufort Sea, although Canadian research on the eastern portion of the Beaufort is helpful. Ironically, it is the region with the least data that is currently of most interest to the oil industry.

Given the size of the regions involved and the relative paucity of physical observations, the information available is marginally adequate for the preparation of EISs. The information base is not adequate to build any detailed predictive circulation model, as would be required to effectively manage an oil spill in the Chukchi and Beaufort lease-sale areas.

Studies of circulation in specific sites identified for production are recommended. This is consistent with the strategy followed by MMS in other OCS areas such as studies of ocean circulation on the Texas-Louisiana shelf and in the Santa Barbara Channel.

Data that have been certified by investigators are available from several federal archives (e.g., the National Oceanographic Data Center (NODC) and the National Snow and Ice Center). In the case of NODC, data gathered on oceanographic cruises funded by the National Science Foundation are required to be reported to and eventually archived at the NODC. Data are also available from researchers in their published results as well as data reports. For example, MMS makes these research and data reports available. An enormous amount of data is available through the EISs.

Physical Properties of Oil and Water

The high viscosity of oil at low temperatures, combined with the influence of surface waves and mesoscale eddies, will lead to spilled oil forming a highly discontinuous pattern of patches, windrows, and sheens rather than a more-or-less continuous slick. Although models for this have been developed and their results are reported in EISs, this topic needs to be kept under review as new information becomes available about the physical properties of crude oil that is discovered in the region and as understanding of the physical oceanic conditions improves.

TRANSMISSION OF NOISE IN THE MARINE ENVIRONMENT

The effect of noise on marine mammals—especially bowhead whales—is a critical issue on the North Slope because Alaska Natives depend on whales for subsistence. Although noise transmission and attenuation in the sea are physical phenomena, this issue is discussed in detail in Chapter 5, where the committee argues that additional physical information alone is unlikely to be instrumental in resolving the issue.

CONCLUSIONS AND RECOMMENDATIONS

The specific conclusions and recommendations for this chapter follow. For the general and overall conclusions, see Chapter 8.

Conclusion 1: The oceanographic model that has been used is elaborate, but inevitably inaccurate because of major uncertainties about the physical processes and the mathematical conditions that are applied to model boundaries in the water. Nevertheless, the model's output is still a useful rough guide to the path and fate of spilled oil, although a simpler model might have been just as useful and credible.

Recommendation 1: *Improve model predictions based on existing knowledge.* Blend in observations of factors, such as the mean circulation in an area, rather than trusting the model to generate this information accurately (which it generally will not). There are numerous ways in which improvement of the model could come from further research on sea ice physics, on interactions of the currents with topographic features, and on the representation or analysis of small-scale turbulence and mesoscale (tens of kilometers and smaller) eddies. Ideally, there would be continuing evolution of the model through comparison of its predictions with new data. More attention should be paid to the tremendous interannual variability in oceanic and ice conditions, and there should be more analysis of extreme events and worst-case scenarios. Further improvements in predicting the fate of spilled oil could come from a better understanding of its spreading and weathering.

Any study of environmental impact is likely to require a model that can predict the path of an oil spill, but after the initial use of such a model it is important to identify important problems about which uncertainty is unacceptably high and for which reduction in uncertainty in the physical oceanographic parts of the problem would be worthwhile. The current approach is "bottom-up," where physical oceanographers strive to produce the best model possible. A "top-down" approach driven by specific environmental concerns and spill-response operations might better focus the research. There is little evidence that this focusing has been attempted, and until it has occurred it is difficult to say with confidence whether the existing environmental information and modeling are adequate.

Alternative: None recommended.

Conclusion 2: Trajectory estimates have relied too heavily on general circulation models (GCMs); they have not been closely tied to observations in the past. The committee also notes that the GCMs for the Alaska region have been prepared by different contractors using different models, with little effort to synthesize the various results or to reconcile the predictions with the limited existing observational data.

Recommendation 2a: *Conduct site-specific circulation studies in areas identified for production and in "hot spots"—breeding, feeding, and aggregation areas—identified by biologists and Alaska Natives with long experience in dealing with the physical environment.* Large-scale studies of the circulation are unlikely to be effective. (The committee estimates that these studies will require at least 2-5 years; they should be subject to continued review. The time required will depend on what the key problems turn out to be and the degree to which interannual variability is a consideration.) In particular, the committee recommends five studies:

(a) Further work on the exchange of water between shallow lagoons and the open sea is required to develop the basic data and understanding that would be necessary to protect them in the event of an approaching oil spill and to facilitate their restoration should they be damaged.

(b) Boundaries between different water masses are frequently zones of significant biological activity and accumulation, and they also can be places where spilled oil converges. A survey of fronts in the Beaufort and Chukchi seas and studies of their causes and variability would be useful.

(c) Interpretation of the ice gouges on the seafloor is hindered by a lack of information about the rate at which they are filled in by the transport of sediment (see Chapter 7). Determination of near-bottom currents (on all time scales) would provide, in combination with measurements of sediment properties, some guidance in interpreting the bottom topography as observed at a specific time.

(d) Long time series of ocean currents and related physical characteristics are required to quantify the interannual and seasonal variability in the lease areas.

(e) Field work and modeling would clarify the extent to which causeways from the shore that have been built or that are proposed in support of industrial operations need to incorporate breaches to permit the continued alongshore flow of water and to permit fish passage. Such studies need not delay leasing decisions.

Alternative: None recommended.

Recommendation 2b: *Refine existing models for use at sites selected for exploration and development and at biological "hot spots." Coordinate modeling with observations in the same area.* Little progress can be expected from refining existing large-scale models. MMS should develop limited-area models for use at sites selected for exploration and development that could be used to explore possible effects on biologically, ecologically, and socially important zones in the vicinity of the production region, to explain how mitigation measures might work, and to predict oil movement should an accident occur. Trajectory models should be used to provide focus and estimates of probability, but they should not be used as the sole determinant of where biological studies will be conducted. If the models are not refined for specific sites, the results will be of dubious value. See detailed recommendations a, b, d, and e in *Recommendation 2*.

Alternative: None recommended.

Conclusion 3: It is well understood that the atmosphere influences ocean circulation in OCS areas, in part because of the local effect of wind stress and heat transport, but also farther away as conditions in distant areas are propagated through the coastal ocean by pressure gradients. In OCS areas off the continental United States and in the Bering Sea, MMS has established a network of meteorological buoys to monitor the lower atmosphere over periods (10 years). There is no comparable set of observations for the Chukchi and Beaufort lease-sale areas, although the committee believes that sufficient information is available from standard numerical weather forecasting products and from fixed-station observations to describe atmospheric variability on scales larger than the mesoscale. The absence of information related to mesoscale variability in the lease-sale areas is not seen as a deficiency at the leasing stage in the existing EIS, but we note that this information will be required once specific production areas are identified.

Recommendation 3: *In production areas and biological "hot-spots," design studies of the spatial structure of the wind field on scales of a few kilometers that emphasize marine boundary layer mechanics and the interaction between the boundary layer and coastal topography interaction.* Such observational studies should involve aircraft surveys and fixed-station observations. To take advantage of the long time series of weather on larger scales, studies also will have to be sponsored that define the relation between the mesoscale variability and fluctuations in the larger-scale weather. The committee understands that the Beaufort and Arctic Storms Experiment (BASE) program of the Canadian Atmospheric Environment Service, which is relevant to this recommendation, is planned for 1994.
    See d in *Recommendation 2*.

Alternative: None recommended.

5 BIOTIC RESOURCES

INTRODUCTION

Mammals, birds, and fish are important to indigenous peoples, and arctic breeding and feeding grounds are essential for the well-being of migratory marine mammal, bird, and fish populations. The potential for damage to living resources is a focus of concern in any plan for the development of oil and gas resources on the OCS. The various acts governing the development of OCS oil and gas mandate that MMS assess the resources at risk and provide estimates of possible effects of OCS activities. The extent of our knowledge of the distribution, abundance, and critical ecological linkages of the arctic biota varies greatly. Some species, especially those held in high public esteem, have been well studied, at least in the summer months, whereas others, usually at lower trophic levels and less glamorous, but perhaps of great importance to food webs, remain virtually unstudied in Alaskan waters. In this chapter, the committee evaluates the adequacy of the biological information available for decisions about OCS oil and gas activities. Where appropriate, we also evaluate the need for additional information. In the committee's taxon-oriented evaluations, it focused on what is known; in the synthesis of its findings, it evaluated the use of knowledge in the production of the various Environmental Impact Statements (EISs) it consulted during its review.

MARINE MAMMALS

The marine mammal fauna of the northern Bering, Chukchi, and Beaufort seas off the coast of Alaska are among the most diverse in the world. Many of the species there are used for subsistence purposes by Alaska Natives and many have an important symbolic role in cultural identity. Some have a high profile because they are covered by international conservation agreements or because they are classified as threatened or endangered under the Endangered Species Act (ESA). All marine mammals in the United States receive special protection under the Marine Mammal Protection Act (MMPA).

The law that protects marine mammals is one of the strongest pieces of environmental legislation in the United States. MMPA was passed to protect marine mammals and to maintain the health and stability of the marine ecosystem. It places a moratorium on the take, including harassment, of all marine mammals with special exemptions for subsistence use by Alaska Natives, for permitted activities such as research and public display, and for restricted permitted take incidental to commercial fishing and industrial activities. Additional protection is afforded to any species that is classified as depleted under the act. ESA requires that any action authorized, funded, or conducted by a federal agency not jeopardize the continued existence of a listed species or stock and not result in adverse modification or destruction of critical habitat. Consultations between involved agencies are required to determine jeopardy, and if jeopardy is determined to exist, then all reasonable and prudent alternatives to an action must be examined. Any species that is classified as threatened or endangered under ESA is automatically classified as depleted under MMPA.

The marine mammals found in the lease areas under consideration include baleen and toothed whales, seals, sea lions, walruses, and polar bears. For some of these species, much or most of their populations spend all or part of the year living in or migrating through the lease areas in the Chukchi and Beaufort seas. Their distribution, movements, and life history events are closely tied to the presence or absence of sea ice (Fay, 1974). Most species are harvested by coastal subsistence hunters, and they can make up a substantial proportion of the annual diet in coastal communities.

Bowhead whales (*Balaena mysticetus*) are an extremely important subsistence resource to Alaska Natives from nine coastal villages, and at those locations they are of key cultural importance as well (Stoker and

Krupnik, 1993). Beluga whales (*Delphinapterus leucas*) are used extensively for food by residents of Kotzebue and Norton sounds, Point Lay, and other coastal villages (Seaman and Burns, 1981; Lowry et al., 1989). Pacific walruses (*Odobenus rosmarus divergens*) are a mainstay of subsistence economies in coastal Bering Sea and Chukchi Sea villages, providing meat, skins, and ivory for handicraft purposes (Fay, 1982). Polar bears (*Ursus maritimus*) and all of the ice-associated seals, but particularly bearded seals (*Erignathus barbatus*) and ringed seals (*Phoca hispida)*, are harvested for their meat and skins (Lentfer, 1988). Shared stocks of polar bears and belugas are also harvested in Canada and Russia, as are walrus, gray whale (*Eschrichtius robustus*), beluga, and seal stocks.

Bowhead and gray whales receive special legal protection under ESA and are listed as endangered. In 1990, Steller sea lions (*Eumatopias jubatus*) were classified as threatened under the ESA. Polar bears also receive special consideration under the International Agreement for the Conservation of Polar Bears, which was ratified in 1976 by Canada, Denmark, Norway, the USSR, and the United States. The Beaufort Sea stock of polar bears is managed under an agreement between user groups in the North Slope Borough of Alaska and the Iñuvialuit Settlement Region of Canada. A joint management agreement for belugas is currently being negotiated between hunters in Alaska and Canada, and walrus conservation and management are being addressed by joint U.S.-Russian working groups. Factors that can affect marine mammals in Alaska are of great concern to Canada and Russia.

Virtually the entire world population of almost 8,000 bowheads and the Bering Sea population of 25,000-30,000 beluga whales winter in the pack ice in the northern Bering Sea, including the Navarin Basin (Burns, 1984; IWC, 1989; Zeh et al., 1993). Bowheads and the Beaufort Sea stock of belugas migrate north through the spring lead system in the eastern Chukchi and Beaufort seas from April until June, en route to their summering grounds in the Canadian Beaufort Sea (Braham et al., 1984). During the autumn migration, bowheads and belugas return through the Alaskan Beaufort Sea and feed along the way. Several thousand beluga whales from the Chukchi Sea stock concentrate near the passes of Kasegaluk Lagoon to molt in late June and July (Frost and Lowry, 1990; Frost et al., 1993). Most gray whales migrate over 3,000 miles from Baja California, Mexico to the northern Bering and Chukchi seas, where they feed during the summer and autumn on extensive beds of benthic amphipods (Jones et al.,

1984). Other cetaceans such as killer whales (*Orcinus orca*), minke whales (*Balaenoptera acutorostrata*), and harbor porpoises (*Phocoena phocoena*) occur in these northern waters, but little is known about their distribution and abundance there.

Four species of ice-associated seals and the Pacific walrus population regularly inhabit the lease areas under consideration (Burns, 1970; 1981a; Fay, 1974; Lentfer, 1988). Ribbon seals (*P. fasciata*) overwinter in the pack ice of the Bering Sea, including the Navarin Basin. A few sightings have been made in the Bering and Chukchi seas in summer, but in general their summer distribution is unknown (Burns, 1981a). Ringed seals bear and raise their pups on stable shorefast ice of the northern Bering, Chukchi, and Beaufort seas (Frost and Lowry, 1981). During summer and autumn they feed in the northern Chukchi and Beaufort seas (Lowry et al., 1980a). Spotted seals (*P. largha*) winter in the Bering Sea ice front, then move north and toward the coast to summer (Shaughnessy and Fay, 1977). Some of the largest concentrations of hauled-out spotted seals in Alaska occur near the passes of Kasegaluk Lagoon along the Chukchi Sea coast (Frost et al., 1993). Summer feeding occurs in the central and southern Chukchi Sea. Bearded seals are present throughout the pack ice of the northern Bering and Chukchi seas during winter and spring (Burns, 1981b). In summer they are found in pack ice over the northern part of the broad Chukchi Sea shelf, where they feed on the rich benthic and epibenthic fauna of this region (Lowry et al., 1980b). The northern Bering Sea and the shallow Chukchi Sea shelf are feeding grounds for almost the entire world population of Pacific walruses. Benthic resources, principally clams and snails, support more than 200,000 walruses (Fay, 1982; Lowry et al., 1980b).

Steller sea lions are widespread in the southern Bering Sea, where they pup and breed on remote rocky shores (Lentfer, 1988). They disperse widely at other times of year, but are almost never seen north of Bering Strait. During fall and winter they can be seen in the Navarin Basin, near sea ice and islands (Brueggeman and Grotefendt, 1984). The importance of this area to feeding animals is unknown. Sea lions occasionally haul out on Hall Island, to the east of the Navarin Basin.

Polar bears den and bear their cubs at coastal sites along the Chukchi and Beaufort seas and on the offshore pack ice. They feed on other marine mammals in the area, principally ringed seals. Two stocks are thought to occur in northern Alaska, one that primarily resides in the Beaufort Sea and northeastern Chukchi Sea, the other in the central and western Chukchi

(Lentfer, 1988). Polar bears are innately curious and can be attracted to human settlements by visual, auditory, and olfactory stimuli (e.g., by movement and noise from vehicles, drilling and other operations, odors from garbage, etc.). Because of this, expanding development of renewable and non-renewable resources in the Arctic has led to increasing interaction between humans and polar bears (Stirling and Calvert, 1983; Stirling, 1983).

Status of Knowledge

Population estimates for gray and bowhead whales are current and updated regularly (IWC, 1990; Zeh et al., 1993). Periodic censuses are conducted for each and results are reported regularly and reviewed widely. Distribution, migration, and feeding are relatively well understood for gray whales (Jones et al., 1984). For bowheads, spring migration through the lead system and summer feeding in the Canadian Beaufort Sea is reasonably well understood. However, once bowheads pass Point Barrow, the autumn westward migration is poorly documented, and the specifics of winter distribution are unknown (Moore and Reeves, 1993). There has been little exchange of information that might be available from Russian scientists about bowheads in the western Chukchi Sea. The importance and regularity of feeding in the Alaskan Beaufort and in the western Chukchi Sea are unclear (NSB, 1987; Lowry, 1993).

For management purposes, beluga whales in Alaska have been assigned to four provisional stocks (Seaman and Burns, 1981; Frost and Lowry, 1990). The validity of these proposed stocks is unknown, and without this information, it is difficult to evaluate the possible consequences of various human activities. Abundance estimates for the Beaufort and Chukchi beluga stocks are based on appropriately designed recent surveys (Frost and Lowry, 1990; L. Harwood, Department of Fisheries and Oceans, Inuvik NW Territories, pers. commun., March 1993), although the estimates are almost certainly not precise enough to measure trends reliably. Information is adequate for identifying migration corridors and important concentration areas during spring and summer (Burns and Seaman, 1986; Frost and Lowry, 1990). Fall migration patterns and wintering areas for belugas are effectively unknown, except that belugas overwinter in the pack ice of the northern Bering and southern Chukchi seas (Brueggeman et al., 1984).

Winter feeding habits are completely unknown (Seaman et al., 1982). Areas that are particularly important for feeding have not been identified. Information on other cetaceans is very limited. Surveys of the Navarin Basin in 1982 and 1983 provided information on species composition and relative abundance (Brueggeman et al., 1984). Belugas, Dall's porpoises (*Phocoenoides dalli*), and killer, fin (*Balaenoptera physalus*), gray, and bowhead whales were present.

Pacific walruses have been surveyed at 5-year intervals under a joint U.S.-USSR agreement. However, interpretation of data is complicated by variable ice conditions, clumped distribution of the animals, and the vast area to be surveyed (Lentfer, 1988; Gilbert, et al. 1992). For the foreseeable future, it is unlikely that survey estimates will be precise enough to detect anything but major changes in the population. Recent satellite tagging and genetics studies have produced useful information about stock separation (Hills, 1992). The Fish and Wildlife Service (FWS), the federal managing agency, has a sampling program to collect and analyze reproductive tracts and stomach contents for comparison with data from earlier studies, but there are no recent data on sex and age composition of the population. These data would probably be the most reliable indicator of status and trend in the populations. Important terrestrial haulouts for walruses and the periods during which they are used are well identified (Fay, 1982). Feeding areas have been generally, but not specifically, identified in the eastern Chukchi Sea.

Information on the status and trends of the four species of ice seals is incomplete and out of date. It will be difficult to predict or manage effects of OCS oil and gas development on these species without any information about whether the populations are healthy or in decline. Spotted seals were counted from 1989 to 1991 at concentration areas along the Chukchi Sea coast (Frost et al., 1993). However, satellite tagging data indicate that these counts substantially underrepresent the number of seals using the area. Little is understood about what causes daily variability in counts. The National Marine Fisheries Service (NMFS) is currently attempting to determine the size of the spotted seal population in Alaska. Ringed seals on the shore-fast ice in northern Alaska were last surveyed between 1985 and 1987 (Frost et al., 1988). Although this study was intended as the start of an MMS monitoring program, no funding has been made available to continue the survey. The most recent counts of bearded and ribbon seals were made in the late 1970s (Braham et al., 1984; Burns et al., 1981). Those data are not only old, but they were not collected as part of a comprehensive and statistically valid survey.

Natural history studies of ice seals were funded by the Bureau of Land Management (BLM), MMS's predecessor, as part of the Outer Continental Shelf Environmental Assessment Program (OCSEAP) program in the late 1970s. However, there has been no recent work to document changes since then. Such studies, as well as updated population information, are particularly needed, because there have been major declines of other Bering Sea pinnipeds during the same time period. Satellite-tagging technology has improved greatly in the past decade. Application of this technique could produce substantial advances in our understanding of the animals' movements, their diving behavior, and their important feeding areas, and it could suggest correction factors necessary for interpreting surveys for all of these seal species. Information on important feeding areas, migratory corridors, and other biologically important areas will be needed to develop mitigating measures to manage the anticipated effects of OCS oil and gas development and production. If such areas are known before development occurs, it will be possible to design activities to minimize harm.

Considerable research has been devoted to Beaufort Sea polar bears (Lentfer, 1988). Population estimates are thought to be adequate or at least as good as current methods allow. There is a reasonable documentation of productivity. Satellite-tagging studies have allowed the identification of important habitats, including denning sites and areas (Amstrup et al., 1986). Similar data for western Chukchi polar bears are more limited (Garner et al., 1990). Studies of Chukchi bears are under way and are being facilitated by the recent ability to do studies in Russian territory, such as at Wrangel Island (G. Garner, FWS, pers. commun., Jan. 21, 1993). Not enough information is available to accurately assess and determine how to mitigate direct and indirect effects of oil and gas activities on polar bears (Lentfer, 1990). Expanding human presence in the Arctic is increasing the potential for bear-human interactions, which may result in the injury and death of both polar bears and people. Oil and gas exploration and development might have adverse effects on polar bears resulting from interactions with humans, such as damage or destruction of essential habitat, contact with and ingestion of oil or other contaminants, harassment by aircraft, ships, or other vehicles, or attraction to or disturbance by industrial noise. While some information is available on reducing bear-human conflicts (Clarkson et al., 1986), more research is needed on possible methods for detecting and deterring bears.

Because they have been listed as threatened under the ESA, considerable research effort is currently being devoted to Steller sea lions, but because the center of their range is well to the south of the lease areas under consideration, they were not a major consideration during this review. Studies of abundance, productivity, energetics, and disease are currently being conducted by federal, state, and university scientists. Weather and ice regimes in Alaska are highly variable from year to year. Because these two factors greatly influence the distribution and movements of marine mammals, there is a great deal of annual variability in when and where animals spend their time. Consequently, it is not possible to conduct short-term studies and consider them representative, or predictive, of long-term distribution or behavior. Furthermore, marine mammal populations are not static. Therefore, no single study of population characteristics or abundance can be adequate. Arctic marine mammal populations should be monitored regularly to detect change and to provide information on annual variability against which to measure change. This need is highlighted by unexplained declines in Steller sea lions and harbor seals (*P. vitulina richardsi*) in subarctic Alaska. One of the clearest lessons to be learned from the *Exxon Valdez* oil spill was that prior information on distribution, abundance, and particularly concentration areas is essential when evaluating the impacts of human activities (Frost et al., 1993). The most fruitful approach is to identify and study "hot spots"—breeding, feeding, and aggregation areas—in the Arctic.

Unresolved Questions

Effects of Industrial Noise

Many marine mammals vocalize, and they rely on sounds in the water for communication and navigation. It is unknown whether exogenous noise might interfere with or mask these functions, or whether it might significantly affect marine mammal distribution and movements. The types of industrial noise introduced into marine mammal habitat could include that from seismic exploration; from barges, transiting supply vessels, and aircraft; and from exploration and development platforms (Richardson and Malme, 1993). Arctic species for which noise is of greatest current concern are bowhead and beluga whales and walruses. The effects of industrial noise have not been clearly revealed by research, despite many complicated and expensive studies funded by MMS and the oil and gas industry. It is possible to argue at great length about the validity of individual studies, but the overriding issue is that there is widespread distrust of the results and dissatisfaction with the design and conduct of studies in arctic communities and others. Unless resolution of this question precedes resource development, it will continue to cause contention between arctic Alaska's residents, MMS, and industry. The question has two aspects: whether noise displaces animals from important feeding areas, concentration areas, or migratory corridors; and whether it makes them less accessible to subsistence hunters. Even if displacement is considered biologically insignificant, it might make successful hunting more difficult. Because the issue is so complicated—compounded by small sample sizes and interannual variability—further studies are unlikely to resolve it soon. Instead, at the end of Chapter 8, the committee suggests an alternative approach that must first involve a critical review of methods and raw data by a diverse panel of qualified acousticians, scientists, and local people who are familiar with the biology of bowhead whales.

Contaminants

Because marine mammals are staple foods in the diet of coastal residents, there is concern about whether they are safe to eat. The baseline data on existing contaminant concentrations in edible marine mammal tissues are inadequate to allow postdevelopment comparisons to be made and reasonably evaluated. MMS has supported the archiving of marine mammal specimens in a national tissue bank, but archiving, not analysis, has been the emphasis of the program. Although the goal of most current development activities is zero discharge of any contaminated materials, there is widespread public concern about contaminant concentrations in animals that are used for food from these industrialized areas. Once industrial activities have begun, contaminants discovered in subsistence foods will be presumed to have originated from those activities. Knowledge of contaminant concentrations in currently harvested resources would provide documentation of concentrations in foods that are now deemed acceptable because they are unaffected by industrial activity. Thus, workup of samples from food harvested in villages would be useful if production and development were likely to occur.

Oil Spills

In an environment where even the simplest human activities can be severely limited by weather and ice conditions, it is likely that a large spill would be difficult and time-consuming to clean up. Such an event and the ensuing cleanup efforts would be likely to affect subsistence hunting activities of coastal residents, as well as the marine mammals they hunt. There is concern in arctic Alaska that a spill would generate concern about bowheads (with or without biological justification) and that after a spill, the International Whaling Commission might reduce the number of whales that subsistence hunters could harvest to ensure that overall mortality did not increase. This could have substantial cultural and nutritional consequences to Alaska Natives.

The effects of oil on most marine mammal species in their natural environments are poorly known. Polar bears, because of their thick fur and grooming behavior, are likely to ingest oil and their ability to thermoregulate is likely to be impaired (Øritsland et al., 1981). Before the *Exxon Valdez* oil spill, effects at the population level had not been observed for cetaceans and pinnipeds, although whales, porpoises, and sea lions have been observed swimming through oil and seals and sea lions have hauled out on contaminated surfaces. Studies following the *Exxon Valdez* oil spill indicated that harbor seals were exposed to and assimilated hydrocarbons and that they suffered nerve damage that likely resulted in death. Based on aerial survey data, investigators concluded that more than 30% of the harbor seals in oiled areas of Prince William Sound died because of the spill (Frost et al., in press).

Cumulative Effects

It is important to note that, with the exception of gray whales, most of the marine mammal species discussed in this report spend much or all of their lives in or near one or more of the lease areas. All of their major life history events, including breeding, bearing young, and feeding, occur at times in areas that could be leased for oil and gas exploration and development. In addition, because many are exposed to other activities, such as commercial fishing and mining, the possibility cannot be ignored that there would be cumulative effects from a combination of events. Although it is unlikely to be feasible to conduct research on cumulative effects, they might have a significant effect on populations and thus it is important to continue to monitor key indicator species, such as belugas.

BIRDS

The Beaufort, Chukchi, and Bering seas are important to migratory birds. Vast numbers of waterfowl and shorebirds nest on the tundra of the North Slope. Several million of these birds migrate in spring and fall along the coasts of northern and western Alaska, and for the populations of many of these species, the concentrations there comprise a substantial portion of their world or North American population. When these birds aggregate in a restricted area, they become particularly vulnerable to severe depletion in the event of an accident, such as an oil spill. Even large, "healthy" populations could be put in jeopardy by a single, severe accident. Species for which a major portion of the North American or world population visits the Chukchi or Beaufort sea coasts include brant (*Branta bernicula*) (Johnson et al., 1992), king eider (*Somateria spectabilis*), Steller's eider (*Polysticta stelleri*), and Ross' gull (*Rhodostethia rosea*) (Johnson and Herter, 1989). The short-tailed albatross (*Diomedia albatrus*), which is endangered, visits the Navarin Basin of the Bering Sea.

Concern for birds that use the marine environment is particularly acute when oil spills are involved. Most species of marine birds spend considerable portions of their time sitting on the water's surface. Most bird species foraging in the marine environment of polar regions repeatedly dive through the water's surface, thus increasing the potential for exposure to oil. An oil spill in ice-choked waters, where the open water is restricted to a few leads or polynyas, could substantially exacerbate the problem by concentrating oil in the limited areas of open water used by the birds. Oiling destroys the water repellency of feathers, on which birds depend for insulation; it also can cause severe physiological pathology when it is ingested by birds attempting to forage or clean themselves (Holmes and Cronshaw, 1977; Hunt, 1987; Nero and Associates, 1987). Most species of seabirds have delayed sexual maturity (e.g., *Laridae* (gulls), *Alcidae* (guillemots, auklets, puffins, etc.)) and low rates of productivity (*Alcidae*). As a result, the replacement of adult birds lost from the breeding population can require decades (Ford et al., 1982; Roseneau and Herter, 1984).

In spring, many birds, in particular the king eider and the common eider (*S. mollissinia*) and, to a lesser extent, the oldsquaw (*Clangula hyemalis*), concentrate in ice-free leads along the coast (Woodby and Divoky, 1982; Roseneau and Herter, 1984). In fall, the coastal lagoons support large numbers of these and other waterfowl including brant, which require safe areas of high productivity in which to molt and forage prior to their southward migrations (Johnson, 1983; Johnson and Richardson, 1981; Johnson et al., 1992). During the fall, large numbers of shorebirds, including abundant red-necked phalaropes (*Phalaropus lobatus*) and red phalaropes (*P. fuli-cari*), also use the littoral zone (and nearshore waters, in the case of the phalaropes) for foraging before they migrate south (Connors et al., 1979; Connors and Connors, 1982; Connors, 1984). In contrast to many areas of Alaska where pelagic environments support the major portion of avian biomass, the most important habitats in the Beaufort for water- and shorebirds are the nearshore, lagoon, and littoral zones. Use of these areas is seasonal, of short duration, and very intense (Connors et al., 1979; Divoky, 1984; Johnson et al., 1992). The ice edge is an area of concentration for ivory gulls (*Pagophyla eburneas*), and a significant portion of the world population of Ross' gulls migrate through the Chukchi and Beaufort seas in the fall (Bailey, 1948; Watson and Divoky, 1972; Divoky, 1983). The Bering and (to a lesser extent) Chukchi seas are the principal foraging areas for millions of short-tailed shearwaters (*Puffinus tenuirostris*), which migrate from colonies in Australia to molt and forage in Alaskan waters between June and October (Roseneau and Herter, 1984; Divoky, 1987; Guzman and Myers, 1987). This species also can be a large contributor to avian biomass in the offshore waters of the Beaufort Sea, where it is associated with intrusions of Bering Sea water (Divoky, 1984). However, overall, the glaucous gull (*Larus hyperboreus*) is the major contributor of avian biomass in offshore waters of the Beaufort Sea (Divoky, 1984).

The Chukchi Sea, eastern Bering Sea, and Bering Strait support some of the largest colonies of marine birds in the northern hemisphere (Sowls et al., 1978). Several colonies consist of 1 million birds or more each (on St. George, St. Lawrence, and Big and Little Diomede islands, for example) (Sowls et al., 1978), and densities of foraging birds near these colonies can exceed 5,000 birds/km^2 (Hunt et al., 1981a, 1990). In contrast, comparatively few seabirds nest along the arctic coasts north and east of Capes Lisburne and Thompson, both of which support large numbers of nesting seabirds (Divoky, 1978b). The lack of significant numbers of nesting seabirds along most of the Chukchi and Beaufort Sea coasts is most likely the result of the lack of nesting sites that afford protection from terrestrial predators such as arctic foxes (*Alopex lagopus*) and grizzly bears (*U. arctos*). Beaufort Sea barrier islands are used as nesting sites by scattered small colonies of black guillemots (*Cepphus grylle*) (Johnson and Herter, 1989). These birds are of interest because they are an outlying population of a species with the center of its distribution in the North Atlantic (Nettleship and Evans, 1985). These populations are also significant because they have been the subject of an unusually long-term population biology study at the University of Alaska. Small numbers of common eider breed on the barrier islands of the Chukchi Sea; most king eiders breed along the North Slope east of Point Barrow and in the western Canadian Arctic (Roseneau and Herter, 1984). Oldsquaw nest along the mainland coasts of the Chukchi and Beaufort seas; relatively few nest on their barrier islands (Roseneau and Herter, 1984).

Status of Knowledge

Colonies at Risk

We now believe we know the distribution and size of most nesting populations of seabirds in arctic Alaska. All major colonies breeding on the Alaska coasts and islands of the Bering, Chukchi, and Beaufort seas have been identified and mapped, and rough estimates of their sizes are available (Sowls et al., 1978). Although in most cases the data are insufficient to provide baselines against which to measure change (or assess the damage from an oil spill), the data are sufficient to determine the approximate sizes of the populations at risk. Data on the distribution and abundance of breeding loons (*Gavia* spp.), grebes (family Podicipedidae), waterfowl, and shorebirds on the North Slope are less precise (for reviews see Roseneau and Herter, 1984; Johnson et al., 1987) but are probably sufficient in most areas for estimating the numbers of birds that could be exposed to an oil spill along the coast of the Beaufort and Chukchi seas, where these birds assemble after breeding (e.g. Johnson, 1983; Johnson et al., 1992). The nesting densities and colony locations are best known for the barrier islands of the Beaufort and Chukchi seas and immediate coastal areas (Divoky, 1978b; Derksen et al., 1981); less is known about the density of nesting waterfowl and shorebirds over the broad areas of tundra inland from the coasts (e.g., data summarized in Johnson and Herter, 1989). Derksen et al. (1981) comment on the potential for harm to water birds breeding on the tundra, and more information on these nesting populations would be appropriate before the development of shore-based infrastructure is approved pursuant to offshore production.

Pelagic Birds at Risk

Information on the pelagic distributions of marine birds in the Bering, Chukchi, and Beaufort seas is relatively scattered and provides less complete coverage than is available for colony sites. There is only a relatively short period during which a large expanse of open water is present in the Beaufort Sea. Few birds are believed to use the open water of the Beaufort Sea (Divoky, 1984) other than the nearshore areas and the ice edge (discussed below). The work of Divoky (1983, 1984) and knowledge of the species of birds that use the Beaufort Sea and those birds' preferences leads us to anticipate that no large or important aggregations of birds occur in the offshore waters of the Beaufort. The data are therefore adequate for leasing and exploration decisions but should be augmented before development and production. It should be noted that the presentation of the data in the two publications cited is insufficient as a basis for judgments about where and when birds are likely to congregate. As a first step, as in any new work, current data should be reanalyzed and mapped. There are few published data on the at-sea distribution of marine birds in the Chukchi Sea during the open-water season, but the unpublished report of Divoky (1987) provides a useful overview and documents a moderately high number of shearwaters and alcids using the central and southern Chukchi Sea. Studies by Piatt et al. (1992), Andrew and Haney (1993), and Schauer (1993) provide a useful overview of the pelagic distribution of birds in the Bering Strait and the southern Chukchi Sea, in which large concentrations of birds are to be expected between June and September. Data for the northern Chukchi Sea are inadequate to assess the potential effects of offshore oil development in that region, and data are lacking on the mechanisms and locations that might lead to predictable, large concentrations of foraging seabirds in the central and eastern Chukchi Sea.

Knowledge of at-sea distributions of birds at sites in the Bering Sea, including the Navarin Basin and regions distant from the proposed lease areas, is important because oil spills resulting from the loss of support vessels or their cargo can damage birds. For the eastern Bering Sea, information on the pelagic distributions of seabirds was summarized in the early 1980s (Hunt et al., 1981b; Gould et al., 1982; Eppley and Hunt, 1984), but numerous studies have been completed since then. New information for distributions in the vicinity of colonies on the Pribilofs (Coyle et al., 1992), St. Matthew Island (Hunt et al., 1988), St. Lawrence Island

(Hunt et al., 1990; Haney, 1991; Elphick and Hunt, 1993)—all in the vicinity of the Navarin Basin—and King Island (Hunt and Harrison, 1990) is available and there are extensive additional unpublished data for seabird distributions around the Pribilof Islands and St. Matthew Island. Consideration of these new data for the Bering Sea will be important for assessing risks from activities of supply vessels.

The nearshore waters of the Bering, Chukchi, and Beaufort seas—particularly the coastal lagoons—are critically important habitats for migrating waterfowl, as are their shorelines for shorebirds. Virtually all studies of these habitats have documented large concentrations of birds conducting the foraging and molting necessary for successful migration (Craig et al., 1984; Divoky, 1984; Johnson et al., 1992). Birds in these areas are particularly vulnerable because they need to acquire food rapidly for successful completion of migration or because they are molting and can be flightless. Contamination of coastal resources by oil spills can affect birds, whether birds are present or not, because the oil can contaminate or kill food organisms such as mysids, euphasiids, and small fish needed by birds during migratory stopovers (Sanders et al., 1980). The oil also can remain in the lagoon and foul birds once they arrive (Humphrey et al., 1990). Although there is considerable knowledge of avian use of lagoons, significant gaps may remain. As recently as 1990, Johnson and colleagues (1992) obtained the first scientific documentation that 40-45% of the North American West Coast population of black brant assemble to stage in Kasegaluk Lagoon on the Chukchi coast before continuing their migration. This is an area of concentration long known to Alaska Natives. Because visits to these lagoons by large numbers of birds are often very brief—but still of critical importance—it is possible that there are other areas, especially along the Chukchi Sea coast, where unrecorded significant concentrations of birds occur. Because the bird aggregations are vulnerable to disturbance as well as to direct oiling, additional studies of remote lagoon systems in the likely path of a possible oil spill are warranted before development begins. Use of the traditional knowledge of Native hunters can facilitate this process.

Sea ice is a conspicuous feature of polar oceans, and one that has considerable influence on the distribution and concentration of seabirds, including Ross' and ivory gulls, which are primarily ice-associated species (Blomquist and Elander, 1981; Divoky, 1984, 1978, 1991; Hunt, 1991). Seabirds tend to concentrate near the edge of the ice and in leads. The annual spring lead that forms between the shorefast ice and the arctic ice pack is a critical source of access to open water and prey for seaducks returning to arctic breeding sites in spring (Roseneau and Herter, 1984). Although published information on avian use of this lead system is sparse, it is sufficient for judgments to be made about the importance of the lead system to birds.

Despite the availability of considerable data on seasonal variation in use of various marine regions by birds, we generally lack information of temporal variability for periods of days to weeks and for periods of more than a year. Given the generally sparse sampling efforts, we do not know what is typical, or what range of variation should be expected. The lack of data on temporal patterns of use is most acute in areas where the major portions of a species' population can occur, and it is important for assessing risks associated with development and production of offshore oil. The recent "discovery" of a previously unrecorded staging ground for black brant is illustrative of the problems of undersampling and of inadequate communication with people who live in and know the area and its resources. Undersampling is a particular problem for the lead systems in early spring and for the ice edge and pack ice in the late summer and early fall. It would be prudent to conduct long-term, periodic monitoring of the distribution and abundance of marine birds and waterfowl in the nearshore waters of the Beaufort and Chukchi seas most likely to be affected by OCS development. A start towards implementing such a program is the MMS-funded development of a monitoring protocol (Johnson and Gazey, 1992).

Processes Affecting Responses of Colonies to Offshore Oil Development

The reproductive ecology and food habits are well known for many species of seabirds that breed in the Bering and Chukchi seas (Hunt et al., 1981a; Springer et al., 1985; Fadely et al., 1989; Mendenhall, 1993), as they are for black guillemots on the barrier islands of the Beaufort Sea (Divoky, 1984; G. J. Divoky, University of Alaska, pers. commun., Aug. 1993). Data also are becoming available for waterfowl and shorebirds that nest on the barrier islands of the Chukchi and Beaufort seas and on the tundra of the North Slope (Johnson and Herter, 1989). These data show when the birds will be present on their breeding grounds or colonies, the kinds of foods they use while raising young, and in some cases, the range of productivity that can be expected. The data are insufficient to allow us to construct accurate life tables, to predict how the birds will respond to major shifts in the environment, or to predict how their population ecology would change after a major loss of individuals. Eventually, we might be able to learn about the responses of various species that lost segments of their populations in the *Exxon Valdez* oil spill, but in many cases the pre-spill (and post-spill) studies have not been sufficiently detailed or of sufficient duration to be very useful. Given our experience with the *Exxon Valdez* oil spill, and given the difficulty of obtaining the data necessary to fully implement the charge to MMS that it must predict environmental impacts and the recovery of ecosystems, the committee acknowledges the impracticality of attempting to gather enough data for MMS to provide more than generalized understanding and predictions of responses of populations to major losses.

Processes at Sea Affecting
Responses to Offshore Oil Development

Process studies of the foraging ecology of birds along the coasts of the Beaufort and Chukchi seas have been more extensive than those devoted to birds foraging in waters outside the barrier islands. Littoral zone foraging by shorebirds has been studied in detail along the Arctic coast and particularly in the Barrow area (Connors et al., 1979, 1981a, b; Connors and Connors, 1982) Likewise, studies of waterfowl use of Simpson Lagoon have provided a good understanding of the critical linkages between shorebirds and waterfowl and their prey in lagoon environments (Johnson and Richardson, 1981; Craig et al., 1984; Johnson, 1984a,b). Surveys of avian use of other lagoons, although less detailed and lacking in integrated process studies, provide sufficient information to let us know that these systems are important for migrating and molting birds and that interruption of the lagoon food webs could displace tens to hundreds of thousands of birds (Divoky, 1978a; Johnson et al., 1992). Connors (1984) discussed interannual variation in breeding success, the number of shorebirds present in the littoral zone, and the timing of migration with respect to environmental variables.

Considerable effort has also been devoted to determining the distribution of birds in the offshore and nearshore waters of the Beaufort Sea (Divoky, 1984). Relatively little information is available on the offshore distribution of seabirds in the northern and eastern Chukchi Sea or on the processes that result in particular areas of concentration. Divoky (1983, 1984) provided information on the average biomass of birds encountered in the Beaufort. We can infer from his examination of avian stomach contents (Divoky, 1976; 1978a; 1984) and from studies in the Canadian Arctic (Bradstreet, 1980; Bradstreet and Cross, 1982) and in the northern Barents Sea (Lønne and Gabrielson, 1992) that arctic cod and various invertebrates associated with the undersurface of the ice (i.e., epontic biota) are the most important sources of food for larids and alcids at sea. We also can infer that the use of the larger epibenthic invertebrates predominates in the littoral and sublittoral zones (Divoky, 1984). Little is published on the at-sea distribution of seabirds in the northern Chukchi Sea or on the processes that determine patterns of distribution and abundance.

Process studies from other areas can be useful for identifying birds in open-water situations where they might concentrate to forage. Piatt et al. (1992) provided information on the diets and foraging ecology of seabirds, particularly auklets, in the Bering Strait and in the southern Chukchi Sea, and they identify processes and the circumstances in which these birds are likely to accumulate and feed. Springer et al. (1979, 1982) provided useful information on foods used by colonial seabirds in the southern Chukchi Sea, but they gave little information on where foraging takes place or about the factors that determine where food will be available.

Knowledge of the distribution and abundance of foraging seabirds in the Bering Sea is important, not only because processes identified there also could function in other regions but also because increased ship traffic in the Bering Sea, in support of Chukchi and Beaufort OCS oil and gas development, might lead to an increased probability of spills in the Bering Sea. Most of these studies have focused on plankton-foraging auklets (*Aethia* spp.) in the northern Bering Sea (Springer and Roseneau, 1985; Springer et al., 1987; Hunt and Harrison, 1990; Hunt et al., 1990), although some work on other alcids and larids at sea is also available (Hunt et al., 1988; Haney, 1991; Elphick and Hunt, 1993). In most instances, the most critical information for predicting where seabirds will aggregate at sea concerns the factors that make prey such as small fish or plankton available. In most of these cases, the behavior of prey in response to physical processes creates regions with concentrations of prey that are especially favorable for foraging birds. These areas can be identified based on properties of the ocean currents, and one can estimate how regularly they will be used by birds. Similar features are likely to be important in areas yet unstudied. Certain biological associations, such as seabird attendance of foraging gray whales in the Chirikof Basin (Harrison, 1979; Obst and Hunt, 1990), are sufficiently consistent that it is possible that foraging gray whales are attended by similar concentrations of birds in the Chukchi Sea.

In contrast, we know little about the physical or biological factors responsible for the observed distribution of short-tailed shearwaters. These birds aggregate in huge flocks, and they are the largest component of seabird biomass in the Bering Sea. Their principal prey, at least as determined in the southeastern Bering Sea, is the euphausiid *Thysanoesa raschii* (Hunt et al., 1981a), but we do not know what determines where and when this euphausiid will be profitably preyed on by shearwaters.

Predicting the Responses of Bird Populations

The immediate response of birds to oiling is well known; most die within a short time, others show effects of varying duration (Hunt, 1987). Much less is known about long-term responses, including rates of population recovery, redistribution of populations, and changes in species composition due to removal of other species. Given appropriate studies, some of these questions could be answered with regard to populations affected by the *Exxon Valdez* spill and by the *T/V Baer* spill January 1993, in the Shetland Islands in northern Scotland. However, for the most part, it is unlikely to prove practical to obtain enough information for precise prediction of long-term responses and recovery times. The task of obtaining the required information is too immense.

Marine birds and others could also react to various forms of disturbance and chronic, sublethal impacts. In the case of disturbance, depending on its severity, birds sometimes desert traditionally used areas, or they can become habituated if the disturbance becomes accepted as nonthreatening (Hunt, 1987; Fjeld et al., 1988). The committee does not know what determines how arctic birds will react to a given level of disturbance either at nesting areas or on feeding grounds. Many site-specific studies have been conducted in the Prudhoe Bay area that have yet to be collated and reviewed as a whole. Such review could point to additional studies on how and to what extent noise, construction activity, or other aspects of infra-structure development and operation will influence birds.

FISHES AND FISHERIES

The fishes of the Beaufort and Chukchi do not form a diverse assemblage; only 72 species have been reported from the northeastern Chukchi Sea (MMS, 1991b) and 62 from the Alaskan Beaufort Sea (MMS, 1990). Of those, a small number of species are of considerable direct and indirect importance to humans. The most important species are members of the salmon and cod families (Salmonidae and Gadidae). The salmon and whitefish (both in the salmon family) species use both fresh and salt water. Species in the salmon subfamily are generally anadromous, spawning in fresh water and migrating to salt water to grow and mature. The arctic members of the whitefish subfamily are amphidromous; they overwinter in fresh water but spawn and mature at sea. The cods are marine and are of key trophic importance in the Arctic. One species, the arctic cod (*Boreogadus saida*), is a major food source for marine mammals and birds, and probably strongly influences their migration (Frost and Lowry, 1981, 1984; Bradstreet et al., 1986). The saffron cod (*Eleginus gracilis*) is also locally abundant and of considerable trophic importance in nearshore waters (Frost and Lowry, 1981). Other species occur in these waters, notably sculpins (family Cottidae) and a few flatfishes (family Pleuronectidae).

Knowledge of these high-latitude fishes derives largely from subsistence fishing and from studies related to oil and gas activities on the arctic coast, including the Canadian Arctic. There is a commercial salmon fishery in Kotzebue Sound (southeastern Chukchi Sea), and some studies were done at the Project Chariot site near Point Hope on the Chukchi coast (Alverson and Wilimovsky, 1966). The most detailed knowledge of the effects of oil and gas activities comes from studies of causeways around the Prudhoe Bay, Kuparuk, and Endicott oil fields in the central Beaufort Sea; the most important fishes potentially at risk are ciscoes and whitefishes (*Coregonus* spp.). Some studies of coastal fishes have been done off the 1002 area of the Arctic National Wildlife Refuge in the eastern Beaufort, and subsistence fishing has yielded local knowledge of ciscoes, whitefishes, and saffron cod.

Many of the studies have focused on nearshore areas (Craig, 1984), although samples have been taken offshore (Frost and Lowry, 1981; Lowry and Frost, 1981). In the Chukchi Sea, additional studies, motivated by attempts to understand the trophic ecology of marine mammals, have also emphasized offshore areas. Kotzebue Sound, a large, protected bay, provides a considerably different habitat from the rest of the more exposed Chukchi coast (Alverson and Wilimovsky, 1966).

The Bering Sea differs considerably from the Chukchi and the Beaufort seas in that it harbors enormous commercial fishery resources, including salmon, herring (*Clupea pallasi*), walleye pollock (*Theragra chalcogramma*), and several flatfish species (MMS, 1991c), in addition to important subsistence fisheries and a small sport fishery. Crustaceans (crabs and shrimp) are also commercially important in this area. Some of the fish and crustacean species in the Bering Sea are important as forage for marine mammals and birds, as they are in the Beaufort and Chukchi seas. Although the Bering Sea contains greater species diversity and biomass than do the Beaufort and Chukchi seas, the current relative lack of interest in leasing in the Bering Sea led the committee to focus on the Beaufort and Chukchi seas. However, if industry interest in the Bering Sea revives, it is likely that the information base there will need careful review.

General Distribution and Ecology

In the Beaufort Sea, the common fish species are Dolly Varden or arctic char (*Salvelinus malma* or *alpinus*),[1] rainbow smelt (*Osmerus mordax*), arctic cod, saffron cod, twohorn sculpin (*Icelus bicornis*) and fourhorn sculpin (*Myoxocephalus quadricornis*), Canadian eelpout (*Lycodes polaris*), arctic flounder (*Pleuronectes glacialis*), and ciscoes and whitefishes (*Coregonus* spp.). Inhabitants of Barrow, Nuiqsut, and Kaktovik use more whitefishes and saffron cod than they do other species. A few pink salmon (*Oncorhynchus gorbuscha*) and chum salmon (*O. keta*) occur in the Beaufort as well (Craig, 1984; MMS, 1990; LGL, 1992).

In the Chukchi Sea, the assemblage of fishes is similar to that in the Beaufort, with some additional species. Salmon, especially chum, are more common in the Chukchi than in the Beaufort, as are char. Starry flounder (*Platichthys stellatus*) and shorthorn sculpin (*M. scorpius*) are common species in the Chukchi Sea. Ciscoes and whitefishes are less abundant and

---

[1] There is not agreement on the taxonomy of the arctic char-Dolly Varden complex of species. The American Fisheries Society's Committee on Names of Fishes (1991) lists them as separate species, with the arctic char (*Salvelinus malma*) being the form found off Alaska's arctic coast, despite earlier works that indicate that the complex should be referred to as *Salvelinus malma*.

diverse than in the Beaufort; saffron cod are more abundant than in the Beaufort (MMS, 1991b). Salmon are abundant in Kotzebue Sound, where they and char are the subject of subsistence fisheries. There is a growing sport fishery for char. Although a considerable amount is known of the life histories of anadromous species (LGL, 1992), less is known about some arctic marine species.

Potential Effects of OCS Oil and Gas Activities

The primary concern in the Beaufort Sea is the possibility of effects on the near-coastal zone, where many fish species feed, migrate, and gain access to freshwater overwintering sites. If arctic cod spawn near or at the shore, then potential effects on them would also be of concern because of their great importance as food for marine mammals and birds (Frost and Lowry, 1981). In the Chukchi Sea, the fishes are probably less susceptible, except in coastal lagoons. Offshore oil and gas activities would probably not affect large sections of a population, and even if they did, their effects could not be measured without spending a very large amount of money and time, if at all. However, an oil spill or other impact might be serious if it affected the local concentrations of arctic cod known to occur offshore and near ice edges (Bradstreet et al., 1986). Therefore, although offshore surveys are valuable for determining the distribution and abundance of marine species and for assessing their roles in marine ecosystems, they probably would not significantly improve our ability to assess or mitigate the effects of OCS activities on marine fish populations off the North Slope or the Chukchi Sea.

Nearshore effects might include those attributable to chemical contaminants introduced by various onshore and offshore activities; alteration of hydrography from structures such as causeways and islands; withdrawal or impoundment of fresh water, making it unavailable for overwintering; contamination caused by abandonment and termination of oil fields; and interference with migration caused by road crossings of streams, causeways, and offshore structures. Under the current regulatory framework, seismic work is unlikely to have significant effects on fishes.

Finally, the problem of tainting should be mentioned. Contamination of the environment with petroleum products or byproducts can taint fish and render them unfit or undesirable for human consumption. In some cases (e.g., the *Exxon Valdez* spill), the perception that fish have been tainted can affect their marketability or availability for subsistence, whether or not they are actually unsafe or unpalatable.

Major Data Gaps

It appears that enough is known about the biology and use by humans to identify possible effects on most fish species. In other cases, even though not enough is known, populations are so widespread that it seems unlikely that a significant proportion of the populations would be affected. In the case of actual development, it would be useful to know more about the effects of terminating and disassembling oilfields. Because of the potential effects on marine fishes and related species, it would be useful to know more about their trophic ecology and life histories, including migration patterns, growth rates, feeding habits, and reproduction, as well as offshore distribution. To obtain such information would require major research efforts, and the costs might not be commensurate with the potential effects on those species. However, because of their trophic importance, knowledge of the spawning habits of arctic cod would be of great interest.

BENTHIC ORGANISMS

Knowledge of the benthos—in terms of its species richness (diversity), standing stock biomass, dynamics and natural history, and capacity to persist in the face of anthropogenic forces or recover from them once the forces are removed—is minimal for the three lease-sale areas. The benthic community in the eastern Chukchi Sea is very rich. Large colonies of benthic amphipods support gray whales. Walruses and bearded seals feed on clams and gastropods. The Beaufort Sea benthic community is restricted to a narrower continental shelf and therefore is less significant to marine mammal populations—very few gray whales and few bearded seals and walruses can be found in the Beaufort.

Zooplankton is a major trophic link. Copepods are one of the main foods of bowheads and arctic cod. Little is known about zooplankton vertical layering and concentrations or interannual variability. Euphausiids and pelagic amphipods (*Parathemisto*) are a major food of ringed seals, arctic cod, seabirds, and bowheads. Mysids are abundant in nearshore waters, which are a major resource for birds and fishes. Little is known about mysid populations' interannual variability and how that variability affects the consumers.

Work conducted in the Simpson Lagoon area has shown it to be an important feeding area for fishes and seabirds. Craig et al. (1984) reviewed work on the trophic dynamics in an arctic lagoon. They reviewed various taxa, especially primary producers, invertebrates, and birds in Simpson Lagoon and compared the overlap in diets of fish and oldsquaw. Primary production was determined by using C_{14} methodology (the results from sampling invertebrates were converted to ash-free grams of dry weight per square meter). Fishes' and birds' diets were surprisingly similar, with both groups being dominated by the "mobile epibenthos," i.e., crustacea and polychaetes. Food chains were short; on average they comprised only three links. Because stomach samples were taken in summer and winter, Craig et al. were able to conclude that there was little annual variation in diet and that mysids and amphipods play a major role in nearshore food webs. They concluded that food supply exceeds demand by 50 times. Even if one makes a more conservative estimate, a food surplus is indicated. The Boulder Patch (Gwydyr Bay) is considered a biologically unique area in the Beaufort because of its unusual flora and fauna, which probably can be attributed to substrate availability. Almost no information is available for the nearshore Chukchi except for an OCSEAP report by Feder et al. (1989). Except for mammal and bird studies, little is known about the major lagoon system at Kasegaluk—very rudimentary fish studies were conducted; no studies were conducted on invertebrates.

In this report, we emphasize the Beaufort and Chukchi seas because the possibility of leasing is greater there. However, the data gaps are a generic problem, as might be anticipated in remote, vast regions characterized by difficult access. Dunton (1992) provided summary biogeographic information on the region's fauna and flora. He identified no references on the capacity of resident benthic assemblages to respond to or recover from imposed stresses. There thus appear to be few, if any, process studies completed in the region. No data are available to assess the long-term effects of sublethal stresses.

Adequacy of the Descriptive Data

Both EISs for Sale Areas 124 and 126 provide references to studies of marine benthos. These are minimally adequate as lists, considering the substantial cost of acquiring further data. They appear to provide little information on spatial variation and no basis for time series analysis of population trends. A recent compilation (Dunton, 1992) lists 51 publications each for the Chukchi and Beaufort seas. Those publications deal with 150 species in the Chukchi Sea and 206 species in the Beaufort Sea. They cover the area from the shore out to a depth of 50 m, and record 13 arctic endemics in the Chukchi Sea and 16 in the Beaufort Sea. It seems important that the EISs identify the endemic species, because their restricted geographic distribution, by definition, renders them more vulnerable. Probably the most significant, yet missing, natural historical detail on any of the resident invertebrate or algal species is how they disperse. High-arctic invertebrates are well known to be characterized by restricted dispersal ability (Thorson, 1950), which bears serious implications for the rate of recovery following local extinction.

Consequences of Proceeding Without Obtaining the Information

Until the problems associated with transportation of oil in ice-filled waters have been addressed and resolved, study of the fate of invertebrate and plant populations should not receive high priority. More information on the topics listed here would be useful for assessing the effects of development and production, especially when information about transportation routes and methods—because that is where most significant spills come from—is available.

• Holes excavated by gray whales and walruses (Kvitek and Oliver, 1986; Oliver et al., 1985) could provide a surrogate for understanding the biotic impact of the deep trenching associated with pipelines buried in the subtidal sediments. Data on the impacts of pipeline trenching are generally unavailable. Without them community response to disturbance and recovery rates cannot be estimated fairly.

• Not enough is known about the importance of the rich estuarine system bordering the Arctic Ocean in terms of primary production and detrital flow to offshore assemblages. In other places, estuarine productivity is important to offshore assemblages. Does this hold for the Arctic Ocean? Is it an ecologically significant factor?

These estuarine assemblages have been minimally examined. The studies of Simpson Lagoon are a valuable exception to this generalization (see Craig et al., 1984). Other coastal lagoons have not been studied or have been studied only minimally. They are certain to be highly significant feeding grounds and migration routes for fish, shorebirds, and waterfowl. Oil spilled in these waters could accumulate on shorelines or be incorporated into sediment deposits and could have serious effects on these systems. Predicting recovery from even a modest spill requires data on sensitivity, reinvasion, and growth rates, at the least. Such data appear to be lacking for most marine arctic ecosystems. However, the Simpson Lagoon (Craig et al., 1984) study suggested that Beaufort Sea lagoons represent an ecosystem subjected to periodic and intense disturbance. Thus, the aquatic plant and invertebrate assemblages might be expected to recover rapidly following anthropogenic disturbance. Such a conclusion is not likely to apply to the extensive bird populations. On the other hand, hydrocarbons can be retained for long periods in lagoons; 19% by volume of oil remained in the sediment eight years after the experimental Baffin Island oil spill (Humphrey et al., 1990). Recovery clearly will not be instantaneous, and its broader ecological and social implications deserve serious consideration.

ALTERNATIVES TO ADDITIONAL STUDIES

There are no alternatives to minimally adequate biological data. For instance, Table 3-4 of a National Research Council report (NRC, 1992a) shows that the fauna at the shelf break in the Beaufort Sea were "not studied," and on the outer shelf they were "not delineated." The Chukchi Sea appears only slightly better studied.

Highly desirable information, such as stock variability and interannual variations in recruitment, probably cannot be acquired quickly. Does this matter or would such information prove useful? The NRC report cited above implies that the answer is either "no" or unknown, even for OCS regions with better characterized benthic assemblages. One conclusion, based on the dominance of largely descriptive studies that place little emphasis on ecologically critical, process-oriented studies, is that "benthic studies . . . have failed to increase our ability to predict impacts on the OCS" (NRC, 1992a, p. 51).

One alternative to further exploratory or descriptive studies would be to examine sites known to have been affected by OCS oil and gas activities. Data on current faunal composition, compiled with information on the dates and areas of disruption and the types of disturbance, could add a valuable perspective on rates of recovery. The lack of this information makes it harder to assess the effects of OCS activities or to predict changes in areas where no activities have yet occurred. At a minimum, the errors of the past should not be repeated; careful monitoring of the effects of OCS activities should be required if any development and production occur.

QUALITY, AVAILABILITY, AND USE OF BIOLOGICAL INFORMATION

In the course of its deliberations, the committee heard several uncomplimentary characterizations of the quality of the science in MMS's environmental studies. The committee did not have the time, the inclination, or the resources to judge the relative merits of consulting firm, academic, industrial, and government science. We found good examples of all of them, as well as examples that might have been better. Instead, the committee provides some guidance here for commonly accepted methods of improving scientific quality. This guidance is provided within the context that the committee judges the biological information generally sufficient to be useful for informing decisions about whether to sell leases in the three areas under consideration here.

Basic Versus Mission-Oriented Science

The most difficult issue is the degree to which MMS should fund basic science as opposed to focusing narrowly on the information it needs for making decisions. The resolution of this issue does not lie in an "either-or" answer. Although it is clear that MMS has a legal and practical mandate to provide information for decision making, it is equally clear that providing this information does not preclude obtaining sound, broadly useful scientific information.

A recent NRC (1993a) report recommended that MMS develop a national framework for its studies and rely much more extensively than it has on the advice of its Scientific Committee for formulating that framework and developing an overall studies plan. If MMS had a broad, national view of the scientific problems it faces and were able to place its regional studies into that context, this committee considers it likely that both the needs of decision makers and the interests of basic science would be better served. For example, a leasing decision in the Beaufort Sea might require information on the birds that occur in the area. A narrow view of MMS's mandate might lead only to a single survey of the birds there. But other lease sales should be anticipated off Alaska's coasts, and some of the birds found there spend parts of their lives in temperate or tropical regions, so broader, longer-term studies would help MMS's decision making and contribute to basic science.

Peer review is widely recognized as an integral part of quality assurance in scientific endeavors. MMS has been requiring that contractors submit papers based on their results to professional journals for peer review and publication. An additional and equally important aspect of quality control is ensuring peer review of the proposed research. The committee is aware of government procurement regulations and of the dangers of awarding grants instead of contracts—scientists have a tendency to follow their interests instead of those of their sponsor if they have grants instead of contracts. Nonetheless, contracts with detailed specifications are not necessarily the best route to obtaining good science, especially if peer review of the proposed research is not included in the process.

MMS has Regional Technical Working Groups (RTWGs) that help to identify important problems for the regional offices. By their composition, the RTWGs ensure at least some community input into decisions about what studies are most important. However, it is not clear that MMS takes sufficient account of indigenous and local knowledge and review by other state or federal agencies, or that the study proposals receive sufficient review by independent scientists. It seems essential that such information be incorporated into the design and interpretation of studies.

Long-Term Studies

Many biological processes—especially those that involve long-lived organisms with long generation times—take many years or decades to be completed and understood. Annual variability in the far north is great and short-term (1- or 2-year) studies often misrepresent or underestimate this variability. Although it is not expected that MMS will solve such problems as understanding the relationship between parent stock and recruitment of fish populations or the determinants of abundance of marine mammals and birds, it is pertinent for MMS to have information on such basic aspects of vertebrate population biology as average generation times, average reproductive rates, food habits, and the variability of their populations over long periods. This should not be interpreted as an endorsement of the attempt to establish a fixed baseline against which any future effects of OCS oil and gas development can be measured. The natural variability in most populations is so great that such an undertaking is unlikely to succeed (NRC, 1978). Instead, long-term and periodic studies can provide information on the range of natural variability and—more important—can provide process information that will be crucial in trying to understand any changes that are observed. Knowledge of what organisms eat, how and when they reproduce, and how long they live is crucial to understanding the effects of disturbances on them and predicting how soon they are likely to recover from those effects. In addition, periodic surveys can provide information on whether animals that breed in colonies return to the same sites in similar numbers each year or vary their choice of breeding sites and colony sizes over some period. Obtaining such long-term information requires a commitment to long-term studies of selected species in selected areas, and it requires some method of ensuring that the results of those studies can be incorporated into decisions and plans for future studies.

The committee notes that the budget available to the Alaska OCS region, and hence its staff, have been reduced in recent years. These reductions will impair MMS's ability to perform long-term research and monitoring and to incorporate the knowledge gained into its planning and into EISs.

SUMMARY

The committee's view of the quality and availability of biological information is summarized in Tables 5-1 through 5-4. Except for Table 5-1, they should not be regarded as final judgments, because it is impossible to know exactly what information is needed until the size, nature, and location of the oil or gas resources are known. In addition, it is important

TABLE 5-1 Status of Knowledge Important for Decision Making (Risk Evaluation) at the Stage of a Lease Sale and Subsequent Exploration

Habitat	Mammals	Birds	Fish	Invertebrates, Plants
Open water	Adequate	Adequate	Adequate	Adequate
Near shore	Adequate	Adequate	Adequate	Adequate
Estuaries, lagoons	Adequate	Adequate	Adequate	Adequate
Leads, polynyas	Adequate	Adequate	Adequate	Adequate
Fast ice	Adequate	Adequate	Adequate	Adequate

TABLE 5-2 Status of Knowledge Important for Decision Making (Risk Evaluation) at the Stage of Production and Related Development

Habitat	Mammals	Birds	Fish	Invertebrates, Plants
Open water	Questions remain	Questions remain	Questions remain	Adequate
Near shore	Questions remain	Adequate	Questions remain	Questions remain
Estuaries, lagoons	Questions remain	Adequate	Questions remain	Questions remain
Leads, polynyas	Questions remain	Questions remain	Questions remain	Adequate
Fast ice	Questions remain	Adequate	Adequate	Adequate

TABLE 5-3   Status of Our Knowledge of Places Where Populations Are Concentrated (Hot Spots)[a]

Habitat	Mammals	Birds	Fish	Invertebrates, Plants
Open water	Questions remain	Questions remain	Questions remain	Questions remain
Near shore	Adequate	Adequate	Adequate	Questions remain
Estuaries, lagoons	Adequate	Adequate	Questions remain	Questions remain
Leads, polynyas	Adequate	Questions remain	Questions remain	Adequate
Fast ice	Adequate	Adequate	Adequate	Adequate

Examples of hot spots: 1) Utukok and Akoliakatat passes for spotted seal hauling out July-Oct; 2) Passes of Kasegaluk lagoon, especially the area off Omalik lagoon, for belugas in late June to mid July; for belugas in late June to mid July; 3) Shorefast ice of Ledyard Bay-highest ringed seal densities; 4) Spring lead system - bowheads, belugas, birds.

TABLE 5-4   Status of Knowledge of How to Do the Remediation Needed If a Resource Is Damaged

Habitat	Mammals	Birds	Fish	Invertebrates, Plants
Open water	Questions remain	Adequate	Not feasible to obtain	Not feasible to obtain
Near shore	Unknown	Questions remain	Questions remain	Questions remain
Estuaries, lagoons	Unknown	Questions remain	Unknown	Questions remain
Leads, polynyas	Not feasible to obtain	Adequate	Not feasible to obtain	Not feasible to obtain
Fast ice	Not feasible to obtain	Adequate	Questions remain	Not feasible to obtain

to recognize that these are the committee's judgments as of the beginning of 1993. As technology, the available information base, and other factors change, our judgments might change as well. Thus, the tables should be regarded as guides to the information the committee believes is likely to be lacking and, in some cases, to the kind of information that might be impossible to obtain for any reasonable expenditure. Because of the uncertainty associated with undiscovered resources, the committee cannot specify the consequences of not having various kinds of information at the development and production stages; those judgments will have to be made when more is known about the extent or resources and development patterns.

In general, the committee concludes that information on biological systems is adequate to make informed decisions about whether to hold lease sales in the areas under consideration. (It is important to note that the committee was not asked to recommend what the decision should be, has not made such recommendations, and that different committee members may believe the information leads to very different conclusions.) For later stages of OCS activities, the committee has grouped the needed information into several categories: adequate; questions remain (information is lacking, but it is possible to indicate what information is needed and how to obtain it); not feasible to obtain (so little is known about the system that the committee cannot identify what needs to be known); and unknowable (either the information is fundamentally impossible to know or the amount of work and time required to develop the information is far beyond any reasonable expectation).

CONCLUSIONS AND RECOMMENDATIONS

The specific conclusions and recommendations for this chapter follow. For the general and overall conclusions of this report, see Chapter 8. In many cases, the committee was unable to estimate time and cost with confidence. Therefore, in those cases, estimates were not provided.

The committee examined the state of knowledge for various groups of organisms in a variety of arctic habitats and concluded that there is adequate information for making informed decisions about the environmental risks attendant in offering OCS lease sales and initial exploratory drilling (see Table 5-2). In making this determination the committee recognized that in the case of biological impacts, there are many risks, and that, even with sometimes sketchy knowledge, bounds could be put on their extent. Whether to accept risk is a policy issue, not a scientific question.

In contrast, the committee concluded that additional information was required before the risks attendant to development and production could be assessed (see Table 5-3). The likelihood of damage is seen as much greater in the development and production stages, and therefore a greater detail of environmental information—often of a site-specific nature—is required for risk assessment. More information about the locations of the major concentrations or points of vulnerability of the populations at risk (see Table 5-4) was seen as essential to the decision-making process at the development and production phases. Specific questions could be identified that the committee deemed answerable and cost effective to answer (such as those pertaining to areas of concentration of birds and mammals in the open water) (see Table 5-4).

In contrast, in most cases, our knowledge is insufficient for remediation or restoration (Table 5-4), and in many instances the committee concluded that the difficulty and expense in trying to obtain sufficient information would be excessive. In this recommendation, we acknowledge that remediation and restoration—part of the task with which Congress charged MMS in its Environmental Studies Program—are in some cases beyond current abilities.

Marine Mammals

Conclusion 1: The issue of the effects of industrial noise on marine mammals, particularly bowhead whales, is unresolved and of significant concern to Alaska Natives and others. Many complicated and expensive studies have been conducted by MMS and industry, but the study design of the studies and the interpretation of the data are highly controversial.

> **Recommendation 1:** *This is an extremely complicated issue that is unlikely to be resolved by scientific study, no matter how much money is spent. An in-depth review of the effects of industrial noise on marine mammals (particularly bowhead whales) should be conducted by non-MMS marine mammal and acoustics experts and should involve a variety of university, agency, industry, and North Slope Borough*

personnel. Experts should review existing reports and data for study design, adequacy of methodology, and validity of interpretation. Subsequent to the review, MMS should encourage involved parties (industry, scientists, and residents) to investigate development of mitigating measures that will address major concerns regarding displacement of seals and whales by industrial noise. Finally, a workshop should be convened to develop recommendations for future studies that will meaningfully address the noise issue. This would require 6 months to 1 year of effort. If no action is taken to mediate conflicts and reach a cooperative solution, it is likely that this issue will continue to disrupt MMS decision making and industrial activities on the North Slope. Millions of dollars could continue to be spent on noise studies that are no more likely to resolve the issue than those already conducted.

Alternative: None recommended.

Conclusion 2: For cultural and biological reasons the highest-profile biological issues associated with OCS oil and gas development involve the bowhead whale. Lack of adequate biological information about bowheads—particularly fall migration routes and the significance of feeding in nearshore Alaskan waters—has led to controversy about when and where OCS activities should occur.

Recommendation 2: *Continue satellite tagging of bowheads to document fall migration routes and to collect data on diving and feeding behavior.* The use of satellite tags can provide a wealth of information on migration routes and speeds, diving and feeding behavior, and to some degree on the bowhead's response to oil and gas activity. This would require 3 years of effort.

Alternative: Delete the bowhead migration corridors from potential lease areas and restrict oil and gas activity to periods when bowheads are not present.

Conclusion 3: There are no monitoring programs for key species of marine mammals in the Beaufort and Chukchi seas lease areas.

Recommendation 3: *Develop a monitoring program for key arctic marine mammal species that reside in areas that will be affected by oil and gas activities. The monitoring programs should be supplemented by stock-identification studies and by satellite-tagging studies of marine mammals to identify important feeding and concentration areas and migratory routes.* Monitoring of ringed seals was begun in 1985-1987 but has not continued. It should be resumed and accompanied by studies to explain the use of habitat and the reasons for annual variations in seal distribution and density. Monitoring should be conducted periodically for beluga whales and spotted seals in the Chukchi Sea. The ringed seal studies should occur for a 3-year span every 6 years and would cost $150,000 per year (including aircraft). The beluga and spotted seal studies should span 2 years every 4-5 years. Without monitoring programs for these species, there is no means of assessing whether significant changes in abundance occur as a result of oil and gas activities. It is also possible that oil and gas activities could be significantly and perhaps unnecessarily restricted if marine mammal population levels are unknown because of federal legislation protecting them. Without monitoring programs, it is likely that population declines will be detected only after they are well advanced, which could precipitate crisis responses by managing agencies and the public.

Alternative: None recommended.

Conclusion 4: There are very few data available on baseline contaminant loads in marine mammals in northern Alaska. It has been repeatedly demonstrated that contamination of food is an issue of great concern to northern residents.

Recommendation 4: *Conduct contaminant studies for arctic marine mammals that are used for food by humans (bowhead and beluga whales, walruses, and ringed and bearded seals).* These studies should be conducted in cooperation with other agencies that conduct continuing studies of some of these species (for example North Slope Borough for bowheads and the Fish and Wildlife Service for walruses).

Alternative: The alternative is to conduct no contaminant studies and to wait for a problem to arise. However, without periodic, continuing analysis of contaminant loads in marine mammals, it will not be possible to assess whether oil and gas activities are to blame. It also is likely that if contaminants are detected in the future and there are no historical data for comparative purposes, oil and gas activities will be considered the cause. This is particularly significant to those who eat marine mammals, because blubber concentrates contaminants at much higher levels than do other tissues and it is preferred for eating.

Conclusion 5: Polar bears are often attracted to industrial facilities in the north, despite considerable care in managing waste disposal and human activities. This can result in injury to humans or the death of bears.

Recommendation 5: *Conduct studies to determine why industrial facilities attract polar bears, and develop and test methods of keeping them away or repelling them once they enter these areas.* There are some promising techniques for deterring polar bears that need further development and testing. This would require 3 seasons of observations.

Alternative: None recommended.

Marine Birds

Conclusion 6: Large numbers of flightless ducks and geese congregate at sea in summer and are vulnerable to spilled oil or other disturbances. Also, it is not known if other seabird species gather predictably in high densities where they might be vulnerable to oil or other disturbance.

Recommendation 6: *Identify the molting areas used subsequent to nesting and determine whether there are open-water areas in the Chukchi Sea (and the Beaufort Sea) where large congregations of eiders and other abundant waterfowl gather.* The areas used and the regularity of their use needs documentation through radiotelemetry and aerial surveys. This would require 3 person-years of effort. Without the information, it will not be possible to protect these resources or provide appropriate mitigation.

Alternative: None recommended.

Conclusion 7: Avian use of the spring lead and polynya system is known to be significant, but the timing and distribution of movement is not well documented.

Recommendation 7: *Determine timing and distribution of avian use of the spring lead and polynya system.* This would require 3 field seasons of effort, aerial surveys, and land-based, small-boat work to determine feeding habits. If the study is not conducted, it might be impossible to provide the appropriate and timely mitigation measures if an accident occurred. We do not know over how broad a front spring migration takes place or the extent to which marine birds are restricted to using any one area of open water.

Alternative: None recommended.

Conclusion 8: In addition to the potential for short-term seasonal and interannual variation in the distribution and abundance of birds, there is the potential for long-term changes in population size, which in some circumstances might threaten the continued existence of a species.

Recommendation 8: *Develop and implement long-term periodic monitoring of the distribution and abundance of seabirds and waterfowl at important colonies and in the nearshore waters of the Beaufort and Chukchi seas.* The monitoring should be at several sites, should occur once every 3-5 years, and should be of sufficient duration at each site in the year of study to cover the entire season of occupancy. This effort would require 2-6 person-years at one to three sites and would last throughout the exploration, development, and production phases. The committee knows of no other method for obtaining a historical data base and for detecting trends in populations that would indicate the gradual development of serious problems.

Alternative: None recommended.

Conclusion 9: If production and development were to occur, spilled oil might affect the estuarine and lagoon systems on which birds depend for part of their food base.

Recommendation 9: *Determine ways to minimize contamination and to restore the food base on which birds depend in nearshore and estuarine and lagoon systems.* Locate areas of avian concentrations and develop plans for protecting these areas from incursion of spilled oil and for remediation should a spill occur. This would require 2 or 3 field seasons of effort plus 1 year for developing mechanisms to protect the most vulnerable sites (location of site with respect to oil-spill probability multiplied by the number of birds using site). Without such studies, it would be more difficult to respond to a spill in a timely and appropriate manner. The fouling of a critical lagoon system could have extremely adverse effects on certain species of North American waterfowl and shorebirds.

Alternative: None recommended.

Fish

Conclusion 10: The available information on fish species in the near-shore areas of the Beaufort Sea is generally good. Information on marine fishes in the Beaufort and Chukchi is limited to very basic distribution studies. With the exception of Kotzebue Sound, little is known about the nearshore fishes in the Chukchi Sea. The information is considered adequate for informed leasing and exploration decisions because species are widely distributed. For decisions about development, production, and termination, more site-specific information on nearshore and fresh-water habitat use by coastal, anadromous, and amphidromous species will be required in most places outside the central Beaufort Sea. For the Bering Sea (Navarin Basin), sufficient information on fish is available to make informed decisions on whether to lease, but careful review of existing information would be needed if a decision were made to proceed with development and production; a considerable amount of additional site-specific information might be needed at such times as well.

Recommendation 10: *Determine whether it is practical to obtain information on the spawning locations of arctic cod in the Beaufort and Chukchi seas. For nearshore anadromous species in the central Beau-*

fort, studies of migration patterns and the effects of offshore structures and causeways should be continued, as should studies of the use of freshwater overwintering areas by various species. These studies should provide a basis for understanding and for conducting studies in other areas.

Alternative: None recommended.

Other Biota (vegetation, benthos, etc.)

Conclusion 11: For marine invertebrates, benthic algae, and especially the estuarine assemblage, responses to pollution of various sorts, maintenance of assemblage integrity, productivity in the face of imposed stress, and potential for recovery subsequent to stress should have been investigated. Questions relating to benthic invertebrates and algae are difficult to answer and must be carefully focused.

Recommendation 11: *Site-specific studies should be initiated as questions arise.*

Alternative: None recommended.

Terrestrial Systems

Conclusion 12: Production and development would result in building onshore facilities to support offshore development. The impacts of the increased infrastructure on biotic resources are not known.

Recommendation 12: *Assess the local and cumulative effects of increased infrastructure development on nesting waterbirds and other wildlife.* Account must be taken of increased disturbance, the loss of breeding habitat, and potential increases in predatory species that would be attracted to the construction of new shore facilities and the additional development likely to follow these new facilities. We do not imply that a new research effort is needed, but cumulative effects need better analysis by all decision makers and EISs should reflect this emphasis. Estimates of the cumulative effects derived from assessment of long-term development scenarios will be useful in assessing alternative strategies for site-specific development. Lessons learned from Prudhoe Bay development should be built in.

Alternative: None recommended.

6 | HUMAN ENVIRONMENT

INTRODUCTION

The Outer Continental Shelf Lands Act (OCSLA) requires MMS to study and manage the effects of OCS activities on the "human, marine, and coastal environments." OCSLA defines the human environment broadly, to include "the physical, social, and economic components, conditions, and factors which interactively determine the state, condition, and quality of living conditions, employment, and health of those affected, directly or indirectly, by activities occurring on the OCS." (43 U.S.C. § 1331(i)). OCSLA thus provides a clear and unambiguous mandate to study and manage broadly defined effects on social systems, as well as on physical and biological systems.

The human environment is affected from at least four causal routes. Two of them are reasonably analogous to the causal routes of effects on the marine and coastal environments. They are actual biological and physical alterations and development-induced changes in the community. The other two, responses to opportunities or threats and long-term or cumulative changes, require more discussion.

Changes in the Physical Environment

Alteration of the human environment that results from disturbances and changes in the physical environment can in turn affect environmental services—the ways in which people use their environment. If an oil spill leads to mortality or contamination of fish populations, for example, commercial, recreational, and subsistence fishing in an area can change. Similarly, ship traffic or pipeline construction can lead to changes in wetlands or shorelines that serve as nursery grounds for fish. Traditional hunting practices in areas that are turned over to pipelines, drilling activities, support services, and other forms of industrial development can be disrupted, curtailed, or eliminated. For example, the migration of bowhead whales, and hence their accessibility to subsistence hunters, might be affected by offshore drilling activities.

Development-Induced Changes

The second causal route involves economic, demographic, and other changes brought about by leasing and development processes themselves. Among its other implications, for example, OCS oil and gas development can lead to increases in population and economic activity, which can have significant implications for the human environment. Some changes are positive, such as the increase in jobs or tax revenues. Others are negative, such as an increase in social pathology due to an erosion of the local culture. OCS oil and gas development can bring additional people to an area—from off-duty workers who choose to live or recreate near where they work, to job seekers, to persons who simply are interested in seeing industrial activity. The increased contact of these new people with Alaska Natives can lead to cultural erosion or to the creation of additional implications for the human environment, as for example through increased hunting pressure on local game.

Responses to Opportunities or Threats

The third avenue of change has less resemblance to those that can be seen in the biological or physical realms. Although the changes to physical or biological systems do not occur until a project leads to physical alterations, observable and measurable alterations in the human environment can take place as soon as there are changes in social and economic conditions, which often occur from the time of the earliest rumors or announcements about a project.

This category involves the patterns of responses that follow the opportunities and threats that attend proposed development. Speculators buy property, economic development opportunities are created, politicians maneuver for position, interest groups form or redirect their energies, stresses can mount, and a variety of other social and economic effects can occur. These can be particularly important for development that produces large changes, such as OCS oil and gas leases; that create controversy; or that are seen as risky by residents. These changes have sometimes been called "predevelopment" or "anticipatory" impacts, but they are real and measurable. Even the earliest acts of speculators, for example, can drive up the real cost of real estate.

Examples of OCS-related effects in this third category—sometimes called "opportunity-threat impacts" in the social science literature (Freudenburg and Gramling, 1992)—include cases in which individuals and communities act to avoid a threat to their futures or to capitalize on potential opportunities resulting from proposed development. These groups also can simply become embroiled in debates over whether a proposed development involves opportunities, threats, or both. Attempts to capitalize on opportunities can include seeking financial gain by anticipating changes in property values or working to gain political advantage from vocal opposition to proposed development. If proposals appear to pose threats, the results can include fear, uncertainty, and doubt about the potential future of the community, which can in turn motivate predictable responses, as when people take time off from work to attend meetings, to organize, or to protest. Many of these important changes in the human environment can take place before a lease sale (NRC, 1989a, 1991a; for a more specific documentation of opportunity-threat impacts, see especially NRC, 1992b, Appendix C).

Opportunity-threat impacts are important not just for purposes of studying and managing OCS activity but also for creating an understanding of and possible resolution to the current deadlock in the OCS leasing process. As a recent NRC review notes, the long-term consequences of this gridlock can include "erosion of public trust in national and local institutions (O'Hare et al., 1983; Baldassare, 1985), alleged economic losses (Cook, 1988), incidents of destructive and criminal behavior (Marshall, 1989), and even large-scale social pathology (Schwartz et al., 1985; Hickman, 1988)" (NRC, 1992b, p. 27).

Social scientists have studied this topic, developed an understanding of the critical issues, and suggested strategies for managing change and resolving conflict (Creighton, 1981; Freudenburg, 1988; Hance et al., 1988; Gregory et al., 1991). Two preconditions for resolving gridlock are defining the problem and developing approaches that acknowledge and incorporate lessons learned. Such strategies by no means guarantee success, but failure to incorporate experience can be a prescription for failure. Overall, as one NRC review concludes,

the public's reactions to government decisions are socioeconomic effects of the decisions and are a legitimate—even essential—subject of socioeconomic study. In addition, a better understanding of these effects could lead MMS to develop a decision-making process that results in fewer and smaller effects than does the current process (NRC, 1992b).

Long-Term, Cumulative Change

The fourth category is long-term or permanent changes to economic, social, and cultural systems that result from the community's response or adaptation to temporary alterations in its environment. As is the case for biological systems, temporary changes to human systems can result in permanent changes in the way these systems operate. Nowhere is this more relevant than in rural Alaska, where indigenous cultures are increasingly integrated with modern industrial society.

Most would agree that dramatic forms of cultural and social disruption should be avoided. Many of the longer-term effects, however, tend to be more subtle, and most are gradual in onset. Because of its remoteness, for example, the North Slope Borough (NSB) already has been identified in the peer-reviewed literature as being particularly susceptible to the phenomenon of "overadaptation" (Kruse et al., 1982; Freudenburg and Gramling, 1992). Local people begin to adapt to changes, but when the changes (extraction of oil, for example) stop, the population cannot return to its earlier condition even if the adaptations are no longer needed or functional. For example, if a hardware store is replaced by a diesel fuel shop, the hardware store may not reopen when development ceases. This is a special problem for the NSB and Northwest Alaska Native Association (NANA) regions because many of the subsistence skills and much of the culture are passed on from generation to generation—a long break could mean that these skills and culture would be lost.

NSB has used money from onshore oil activities to build an infrastructure that requires substantial maintenance. As the oil money declines and finally runs out, maintenance could become a significant burden on the community. It is unclear whether the Iñupiat economy and culture will be able to assume that burden and survive.

One ironic result of what often happens when dependence on extraction (dependence on revenue from oil and gas development and production) is coupled with remoteness is that there can be a long-term increase, rather than decrease, in local poverty. This is contrary to common expectation, and it happens despite high average wages—not only after the shut-down of operations, but often even during the operating lifetimes of extractive facilities (Drielsma, 1984; Elo and Beale, 1985; Tickamyer and Tickamyer, 1988). Although such a result can be predicted, it is not often adequately anticipated; the necessary studies are not conducted in advance. MMS has not conducted studies that would identify ways of mitigating or managing boom-bust impacts. Further discussions of boom and bust resulting from oil and other resource development are found in the work of Bunker (1984), Cummings and Schulze (1978), Gulliford (1989), Krannich and Luloff (1991), and Weber and Howell (1982).

DISTINCTIVENESS OF THE HIGH ARCTIC AND ITS PEOPLES

One distinctive characteristic of the North Slope, as well as of other areas of rural Alaska, including the Kotzebue-NANA region, has to do with the continued vitality of local tradition. The North Slope is home to several thousand Alaska Natives, whose traditions in many respects are quite strong. The Iñupiat people and culture are unique in that they are the U.S. inheritors of the only cultural tradition to survive and thrive in high arctic conditions, although they resemble other indigenous U.S. cultures in their strong adherence to traditional activities.

Because of the cold and because of the radical annual cycles of daylight and darkness, the high arctic is the most extreme environment to which the human species has ever successfully adapted. The human adaptation has developed over 4,000 years in the area from Greenland to the Bering Strait. Within that region lives a relatively uniform population—the Iñuit or Iñupiat—whose unity is in part reflected by similar linguistic and genetic characteristics. The traditional high arctic populations of the Beaufort and

Chukchi seas represent one regional variant of culture adapted to the extremes of cold weather, lack of sunlight, and highly seasonal patterns of resource availability. Studies of the biological characteristics of Alaskan Iñupiat suggest that they have developed several genetic adaptations to conditions of extreme cold and food shortage not found in other areas of the world (Jamison et al., 1978; Moran, 1982). The Alaskan Iñupiat also have adapted or attempted to adapt to a variety of radically new and different ways over the past century and a half. Since the 1860s, when the first Yankee whalers passed through the Bering Strait in pursuit of the bowhead whales, adjustment to Euro-american ways and doings have been a part of Iñupiat cultural heritage. Contemporary Iñupiat behavior and cultural practices are a complex distillation of technologies, genetics, behavior, and institutions. Through it all, however, core features of Iñupiat identity and cultural practice have been retained and passed on.

Iñupiat Culture

Culture is perhaps the most uniquely human of characteristics; all human beings are raised in cultural systems, which include not only a way of life, but also a reasonably coherent understanding of the meaning of life and of our own place within it. The resultant sense of coherence and integration of ourselves within a larger context is also what gives us a sense of purpose and self-esteem.

All human societies transmit cultural systems to their youth, but each generation responds as well to environmental changes, often making new judgments about who they are and how they wish to live, resulting in both continuity and change across generations. The issue is not whether cultural change will occur; it always has and always will. The questions are, instead, what will be the nature of that change and to what extent will the affected people be able to decide their own futures? The answers to these questions can exert a powerful influence over the degree to which the resultant change will, on balance, provide for fulfilling lives or lead to cultural disruption and social pathology. This is the foundation of the OCSLA definition of the human environment as including "quality of living conditions" (43 U.S.C. § 1331 (i)). Evidence indicates that the results of change can include severe trauma and stress, as well as confusion and breakdown in cultural transmission from adults to children. Studies of the

Exxon Valdez oil spill (Impact Assessment, 1990; Palinkas, 1990), of population relocation caused by dam building in Egypt and other parts of Africa (Colson, 1971; Scudder, 1973; Scudder and Colson, 1972), and of political displacement in Arizona (Scudder, 1982) demonstrate that serious disruptions to the continuity of cultural systems that are tied to specific locales can result in a declining quality of life (as measured by the poor health of individuals, the dissolution of social relations, and the disappearance of cultural institutions).

For Alaska Natives, as for any other culture, change is constant, and even within a relatively small community, it is possible to discern a continuum, from those who are extremely traditional to those who are most similar to the nonnatives of the lower 48 states. Even in the case of such traditional activities as whaling, the crews tend to use some modern equipment (motorized aluminum skiffs, used primarily in the fall season) along with more traditional equipment (sealskin boats, used primarily in spring whaling). In many ways, this approach to the adoption of nontraditional technology is similar to what can be found among midwestern farmers, for example, who use modern tractors rather than horse-drawn plows and yet who continue to see themselves as part of the long-established traditional activity of farming. Through their creative blending of traditional and modern ways, the North Slope Iñupiat have maintained a remarkable degree of cultural integrity and community cohesion.

Core elements of the Iñupiat culture, upon which their distinctive and valued identity are founded, include subsistence, ceremonial activities, language and music, and kinship and family relationships.

Subsistence

Distribution of subsistence hunters is fairly well known and is extensive due to modern small-scale motorized transport. Subsistence activities include hunting, capturing, processing, distribution, and consumption of wild animals, especially the bowhead whale. Meat and fish consumed by Arctic communities come from marine and terrestrial harvests in widely varying proportions from year to year (S. Pedersen, Alaska Department of Fish and Game, pers. commun., Nov. 1993). The bowhead whale is central to Iñupiat self-perception and has become increasingly important in the past 15 years as a symbol in external and international cultural arenas. The bowhead whale hunt mobilizes a large portion of the population. Adult men lead and manage as *umealiks* (captains of the hunt), coordinating their crews into hunting units. Younger men participate as crewmen and in the cutting up of the animals. Children and adults alike eat *maktak* (skin and blubber) as it is cut from the whale. Women keep knives and *ulus* (traditional skinning knife) sharpened and collect the portions they take home to their families.

Ceremonial Activities

Tightly linked to subsistence practices and perhaps a crucial element in the maintenance of a collective, public identity for the Iñupiat is the regular practice of religious and cultural ceremonies. Their link to subsistence comes from the centrality of giving gifts of subsistence products to others as a demonstration of generosity and goodwill. These ceremonies include games, dancing, singing, and gift giving.

Language

Speaking Iñupiaq is still a strongly valued cultural characteristic. Many adults and elders continue to speak it as a first language and prefer its use on occasions of public testimony. The language is particularly important because of its vocabulary for identifying environmental conditions of ice and snow as well as the characteristics of animals and their behavior (Nelson, 1969). Although television and formal education have promoted the use of English among younger generations, there continue to be many who understand and attempt to use Iñupiaq. The language is now taught in the schools in the hope of continuing its use.

Kinship and Family

Sharing and interacting among extended family continue to be important components of Iñupiat culture (Luton, 1985; Jorgenson, 1990), whose practices include adoption, name sharing, and resource sharing. Occasions to celebrate family ties often derive from the distribution and consumption of subsistence products, which are crucial both culturally and as a source of nutrition for many families because of the extraordinary cost of imported foods, which often are inferior nutritional quality. Kinship is transgenerational in that the Iñupiat believe that the spirits of the deceased reappear in newborns who are named for their ancestors.

Extreme Isolation

Because of the remoteness of the North Slope and western Alaska from potential markets and supply sources, the region faces understandable but considerable challenge in attempting to compete, in most industries, with locations that support larger populations and have a climate that is less severe. As already noted, the North Slope Borough presents the most extreme example in the United States of the challenges of geographic isolation. Hundreds of miles separate the borough from Fairbanks, the nearest metropolitan area; basic food products, such as agricultural plants and livestock, are virtually impossible to keep alive outdoors. Even more so than for the more southerly regions of Alaska, the North Slope has a high degree of dependence on economic input from the lower 48 states, exacerbated by the high cost of transporting even basic goods, such as fruit and vegetables, to a region that has little capacity to grow its own. The region also has an extremely high dependence on extractive industries and hence especially high levels of susceptibility to potential boom-and-bust disruptions.

Experience shows, moreover, that these disruptions can include significant problems related to busts even if the boom periods are managed reasonably successfully. These conditions present challenges for many, if not all, of the potential antidotes that are prescribed for dealing with similar disruptions in less remote regions. Perhaps because of a previous failure to recognize such impediments to diversification, until the past few decades it was relatively common to encounter references to continued adherence to subsistence activities as reflecting an unfortunate weakness of economic rationality among indigenous peoples. As noted by Bowles (1981), among others, closer attention to the facts of extractive activities in remote locations tends to point to a different conclusion. In all too many cases, even what appeared once to be steady and reliable jobs in mining or oil extraction will be found within a few years to have succumbed to the notorious volatility of world commodity markets. In short, even these steady jobs can disappear just as quickly, as did thousands of oil-related jobs in southern

Louisiana during the 1980s (Centaur Associates, 1986). Subsistence ways of life, by contrast, display remarkable resilience over the centuries, often in locations that seem largely unsuited for formal economic activities, other than the extraction of raw- material deposits that happen to be both large and unusually rich. In light of experience, a continued commitment to a subsistence economy can be a highly rational form of insurance against the inherently unpredictable fortunes of extraction.

North Slope Borough

Rapid and radical changes in the past 20 years have dramatically transformed the living and working conditions of the North Slope Iñupiat. The creation of the North Slope Borough has been a major factor in buttressing Iñupiat cultural practices and in buffering the people themselves from participation in the oil industry labor force at Prudhoe Bay.

Before 1970, cash was relatively scarce on the North Slope. There were low-paying local government jobs, but the money came mostly from the sale of furs or handicrafts and from seasonal construction employment, which often was available only at a distance from the home community (Chance, 1990). Living conditions were harsh; the one- and two-room tarpaper shacks of most residents lacked running water or sewage treatment facilities. Tuberculosis and bronchial problems were endemic (Chance, 1990). Because they were forced to enroll in high schools in the lower 48 states and serve in the U.S. military, many Iñupiat were introduced to the standard of living and quality of life available in most other parts of the United States. In the face of some emigration in search of employment and a more comfortable life, the Iñupiat leadership determined to create a better quality of life on the North Slope. When this was accomplished, many displaced Iñupiat returned home.

The discovery of oil at Prudhoe Bay and its eventual commercialization through the Trans-Alaska Pipeline System in 1979 opened up an entirely new era in North Slope Iñupiat existence. The development was perceived by the Iñupiat as having the potential to provide the improvement in the quality of life desired by most North Slope residents (McBeath, 1981). Under the leadership of Eben Hopson, in 1972 the elders formed a borough (now the North Slope Borough), which would encompass the Prudhoe Bay field (McBeath, 1981; Chance, 1990).

Through this instrument, taxes were levied on oil industry facilities and infrastructure at Prudhoe Bay; the revenues were used to underwrite the creation of a government bureaucracy and to fund the Capital Improvement Project (CIP). Through the CIP, borough tax revenues derived from Prudhoe Bay oil-field facilities were used to provide construction jobs and improved living conditions. Substantial investment in all North Slope communities was made in new, upgraded housing; in freshwater and sewage treatment systems; in airport runways; in schools; and in government buildings. Iñupiat men went to work for substantial wages and thereby upgraded their subsistence equipment and household inventories (Kruse et al., 1981; Jorgenson, 1990). In addition, the borough government created a substantial number of clerical and administrative positions, held mostly by women, providing relatively stable cash incomes for many households. The North Slope Borough was politically controlled by the Iñupiat, who elected a mayor and assembly members. Although this arena became subject to substantial competition among Iñupiat families and factions, it provided a buffered, insulated opportunity for the Iñupiat people to obtain employment and income. Thus, over the period of the 1980s, very few North Slope Iñupiat took permanent employment in the oil industry. Most preferred some form of temporary or part-time employment that was more compatible with subsistence hunting. The borough's hiring and employment practices buttressed this culturally preferred pattern (TR 125).

The North Slope Borough also has promoted the preservation and retention of Iñupiat cultural knowledge and tradition. Annual elders conferences have been funded, and biographies and information about the history and practices of the Iñupiat have been recorded. An Office of History and Cultural Heritage has been created to oversee the collection, storage, and use of this material.

Through the creation and control of the North Slope Borough, the Iñupiat obtained revenues for use in ways that are congruent with Iñupiat values. Because the North Slope Borough is a political arm of the state of Alaska, its practices are subject to state law, and control of its power can be maintained only through democratically elected officials. The substantial Iñupiat majority has dwindled as newcomers have been attracted by the successful economy. If in the future the direction of the North Slope Borough were to pass out of Iñupiat hands, current policies and practices could change.

SOCIAL AND ECONOMIC STUDIES

From its inception in 1977, the Alaska regional office of MMS has conducted a range of social, economic, and cultural studies as part of the planning process for OCS leasing and development. These have included baseline studies of economic, demographic, and social conditions; community studies; harvest disruption and impact studies; and special studies on institutional characteristics, social indicators, and employment (currently under way). Slightly less than $26 million has funded MMS's socioeconomic research in Alaska, which is about 6% of its total Environmental Studies Program budget for the state.

This section describes the Socioeconomic Studies Program (SESP) and discusses the extent to which MMS's analyses provide coverage of baseline conditions and the various potential effects on the human environment that could result in changes to those conditions. The focus is on MMS's studies, because most of the site-specific information about the three lease-sale areas has been collected by MMS, and most of the information gaps will most likely have to be addressed by MMS rather than by others.

Baseline Studies

Baseline studies of the demographic, economic, social, and cultural characteristics of the North Slope region were conducted primarily from 1978 to 1986. Information in those studies was based on 1980 census information and on census updates from the North Slope Borough. The committee was not made aware of any studies based on the 1990 census.

Community-specific, baseline descriptive studies of a quasi-ethnographic nature have been used to develop cultural profiles through the anthropological methods of field work and participant observation. Normally, this requires an extended period of residence—a full year is considered standard for this kind of research. The ethnographic description of the community identifies the nature of cultural behavior observed, ranging from subsistence and economic behavior through kinship and family behavior, ceremonies, religion, and the expression of cultural values in a range of contexts. MMS community studies typically have involved residence by an ethnographic field worker, on an intermittent basis, for 6-9 months. The descriptions provided by these studies have tended to focus on modal or "typical" behaviors, presenting these as the cultural expression of the community. They do not provide information on the range of variation in subsistence hunting, ceremonial participation, kinship relationships, and the like. SESP has done a credible job of recording baseline conditions on the North Slope of Alaska. The studies provide a detailed and knowledgeable description of current conditions and discussions of how those conditions are likely to change in the near future. Although there are weaknesses in the coverage of non-Native populations—including consideration of the relationships between Alaska Natives and non-Native populations in the North Slope and elsewhere—MMS studies have contributed significantly to our understanding of the performance of the social and economic systems on the North Slope. Such studies are required if MMS is to assess the impact of OCS oil and gas exploration and development realistically in light of recent population increases in North Slope communities.

Subsistence

Subsistence was the focus of extensive research on the North Slope Borough from 1987 to 1990 to document the areas, seasons, and harvests of three major communities. The methods used focused on the most active participants in subsistence harvesting in the communities, who were asked to report on their activities. It appears that the studies represent an adequate baseline for at least initially determining community harvest levels and typical and preferred hunting areas.

The effects of harvest disruption in the Chukchi Sea area have been examined, focusing primarily on marine mammals. No disruption or impact studies have been done for the Beaufort Sea region.

Physical Disturbances

Disturbances of the physical environment and the changes they bring have received a significant amount of attention, at least at an initial level of analysis. MMS studies describe, often in considerable detail, the amount and kind of subsistence activities, the importance of subsistence activities for the maintenance of traditional cultures, and at least the potential for these activities to be disrupted in the case of catastrophic damage to the physical environment. Less attention has been devoted to the ways in which subsistence activities, and the broader cultural significance of given environmental settings, might be disturbed or affected by OCS oil and gas activities even in the absence of major spills. This weakness, in turn, leads to neglect of the pragmatic steps that might be available for avoiding or mitigating such effects, examples of which range from the outright loss of certain areas for subsistence purposes because of their conversion to industrial use to the loss of areas for pipelines and processing facilities. Less overt problems also merit study: These include noise, which is purported to drive some species from traditional seasonal use areas, and the increased survival of human populations, which puts more pressure on wildlife.

Economic Growth

Considerable attention has been paid to the economic impacts of OCS oil development through MMS's development of economic growth models and related work. Economic growth models simulate economic, fiscal, and demographic changes caused by various forms of development. First, development scenarios are prepared that specify the number, kind, and probable locations of oil and gas development facilities. These are translated into employment estimates, which are used to project socioeconomic conditions. The economic growth model has three integrated components: economic, fiscal, and demographic. The economic component projects employment, wages, real disposable income, prices, and total output. The fiscal component projects the level of government activity, such as personnel expenditures, state government employment, and expenditures on capital improvements. The demographic component projects population change through separate projections of three components: births, deaths, and migration. The models are then used to project changes that could result in these demographic and economic factors from OCS development. The projections from MMS models appear to be consistent with the general state of practice in other locations.

Mixed Cash-Subsistence Economies

The status of economic activity within the North Slope Borough is reviewed in several studies that provide significant insight about the evolution of this remote society over the past two decades. The studies show that the Iñupiat have demonstrated a remarkable ability to develop and adapt to new institutions and to use them to promote their political and economic welfare (TR 125). The dominant financial effect of oil development on the local communities was the influx of cash through property taxes on oil facilities in Prudhoe Bay. The funds collected were used to finance capital improvements that resulted in large numbers of jobs for local residents. Iñupiat men took the majority of high-paying, temporary, construction-related jobs; they then became the largest unemployed sector of the population as construction declined (TR 125). Iñupiat women tend to hold lower-paying, but permanent, jobs in administration and to achieve leadership positions in new institutions (TR 125).

Direct employment of Alaska Natives by the oil industry has been far less significant (TR 85; TR 120). The major reasons for this appear to be an absence of formal training and certification—Alaska Natives often are not adequately trained or certified for jobs in the oil industry, such as welding or heavy equipment operation—as well as an unwillingness of Iñupiat men to commit to steady shift work if it conflicts with hunting or village activities (TR 85). MMS found that many Iñupiat view oil industry jobs unfavorably (TR 85). Increased government revenues within communities stimulated rapid business expansion; however, most new businesses and professional and high-skill positions are largely dominated by non-Natives (TR 117).

Subsistence is an important component of the mixed economy (TR 133), despite the rapid growth in the cash sector. In many cases, proceeds from the cash economy served to finance subsistence activities (TR 125; TR 133) rather than to substitute market commodities for subsistence harvests. Indeed, many of those with jobs in the cash sector view employment as a temporary effort to collect enough money to finance subsistence activities. Also, many wage earners continue with subsistence activities in their spare time. Cash obtained from jobs is used to continue the Iñupiat traditions of sharing, whereby employed individuals—particularly the elderly—provide cash to hunters, who return a share of the harvest (TR 125).

To date, conflict between oil development and subsistence activities has been isolated; the major impact has been in regulatory restrictions on subsistence land use in development areas (TR 85). However, there is an intense and widespread concern among the Iñupiat that offshore development will harm subsistence hunting, and this could lead to significant social stress (TR 85).

The studies conclude that there will be an eventual decline in the cash economy as oil revenues decline. The proposed OCS activity could help forestall the decline in borough revenues and could stabilize Alaska Native populations for more than a generation (TR 100), but the lease-sales will not bring enough income to reverse the decline in tax revenues, expenditures, employment, or population (TR 120).

Developmental Activities

The economic effects of development also have received significant attention. Social and cultural effects have been less thoroughly studied. Experience to date shows that Alaska Natives received very few of the jobs available at the major North Slope oil operations—the North Slope Borough estimates less than 1%. On the other hand, there are several oil-field service companies (for catering, etc.) that are owned by Alaska Native corporations. These appear to compete successfully for business and to hire a high proportion of Iñupiat workers. In addition, the North Slope Borough, in particular, experienced considerable success in using oil-derived tax revenues to support the hiring of Alaska Natives who work directly for the borough and, hence, indirectly for one another.

SESP contributed significantly to the decisions to pursue an "enclave" approach to development, lessening significantly the amount of interaction and potential conflict between oil workers and Iñupiat residents. This approach appears to be seen as having achieved reasonable success. It does not appear, however, that SESP has dealt as successfully with anticipating, quantifying, or otherwise documenting the broader range of social and cultural effects directly associated with development—potentially ranging from easily quantified measures, such as crime rates (Freudenburg and Jones, 1992) and the proportions of non-Native residents in the North Slope Borough, to less easily quantified, but arguably no less significant considerations, such as the degree to which a higher proportion of non-Native residents would be expected to lead to a loss of cultural pride and the sense of Iñupiat self-determination, or about steps that could be taken to manage or mitigate such changes.

Social Indicators

The basic logic of the study of social indicators is most often expressed by analogy to economic indicators. Just as some economic statistics—such as the rate of inflation or the measures of productivity—can be seen as "indicating" the overall health of the economy, social indicators are intended to provide documentation on the overall level of sociocultural health in a community or region. MMS began its social indicators studies in 1987, and the effort has since received a significant fraction of the overall funding for socioeconomic studies in the Alaska region.

Given the special requirements of the Alaska region, SESP had to modify the methods and goals that are standard in other social indicators studies. In many ways, the modifications resulted in studies that complement MMS's economic modeling effort. The social indicators studies display a keen awareness of the special nature of the Alaskan context, particularly with respect to Iñupiat villages. The studies produced rigorous estimates, for example, of the degree and kind of sharing of subsistence take, producing information that could provide a useful baseline against which future changes could be compared. At the same time, the studies do not appear to have provided the detail or kinds of information that would normally be expected in a state-of-the-art social indicators study. There appear to be gaps in the kinds of data gathered from government bodies (on physical and mental health and on "social pathology" indicators, such as crime and violence, drug and alcohol abuse) and in the kinds of data normally gathered through standard survey techniques, such as individual assessments of well-being, community stability, and quality of life.

Opportunities and Threats

MMS-funded studies do not appear to have dealt, except in a highly preliminary way, with the responses to opportunities and threats that can be expected to result from proposed development. To MMS's credit, several studies have noted that the world views that are taken for granted within the agency are not necessarily shared within Iñupiat communities, and vice versa, and several studies have taken pains to point out that neither set of views should be considered superior. To date, however, the studies do not appear to have considered the likely social and economic responses to opportunities and threats—the more specific patterns of behavior likely to result from the opportunities and threats that are perceived by affected populations, including the responses of non-Native as well as Iñupiat populations. Even though the information on Alaska Native world views can at least formally be considered to be "available" for decision making, it is impossible for the committee to judge the degree to which it has been taken into account.

Based on existing social science literature, it is clear that opportunity-threat impacts (those anticipatory impacts that result from people's seeing an impending physical or economic change as an opportunity or threat (Freudenburg and Gramling, 1992)) will occur if leasing goes forward, and the general nature of those impacts can be anticipated. The social science literature also makes clear that opportunity-threat impacts can be mitigated, at least in part, by establishing an appropriate decision-making process. Further discussion of mitigating opportunity-threat impacts in presented in Chapter 7.

Although the social science literature provides general lessons, opportunity-threat impacts often vary in ways that are highly case-specific, particularly for communities and cultures as unique as those in Alaska. Thus, no amount of general study will provide the kind of site-specific information that is required to understand and manage the effects of any particular lease sale. Given that little attention has been focused on opportunity-threat impacts in northern Alaska, little is known about the specific form and severity of the impacts that can be expected for leases in this region. Also, little is currently known about the appropriate pragmatic steps that could be taken to mitigate those impacts within this particular context.

In addition to the need for an improved understanding of opportunity-threat impacts in the region as a whole, site-specific assessments of potential impacts will be needed before it is possible to make informed decisions about specific lease sales. For these assessments to be taken seriously, they will need to be considered in adequate depth and detail in the decision-making documents related to a given lease sale.

Longer-Term Adaptations

The committee has been unable to find evidence that SESP has dealt in a systematic manner with long-term or permanent socioeconomic changes that are likely to result from OCS development. Several studies do refer generically to the potential for an eventual "bust" once the petroleum reserves are exhausted (TR 125), but they have not dealt adequately with the more gradual (and equally predictable) problems of cultural erosion and overadaptation. Some evidence of the potential for problems from development at Prudhoe Bay already can be seen. The North Slope can provide a textbook illustration of the potential for overadaptation, but it does not appear that systematic studies have yet focused on this potential or on the steps that could be taken to deal with the likely result. In addition, far too little information is available about specific implications of extractive dependence for the North Slope Borough.

CONCLUSIONS AND RECOMMENDATIONS

The specific conclusions and recommendations for this chapter follow. For the general and overall conclusions of this report, see Chapter 8.

MMS's socioeconomic studies in Alaska deserve considerable credit for the scientific progress they reflect. The Alaska SESP conducts extensive and substantive reviews of the social and economic impacts of OCS activities on the North Slope of Alaska. The quality of work, judged on its own terms, is generally quite good.

SESP has established a credible baseline analysis of social and economic conditions in Northern Alaska, and a number of important questions have been addressed. Furthermore, the program has carried out some studies that analyze potential changes, particularly economic ones, associated with OCS oil and gas development. That work, however, has failed to deal adequately with other issues that are critical to projecting and managing social change. As a result, a significant fraction of the information that is necessary for informed decision making is unavailable, and as a corollary, not enough effort has been devoted to the pragmatic questions of what steps, if any, could be taken to avoid or lessen effects.

Although it is important not to underestimate the value of the solid foundation that has already been provided by SESP, it is also important to recognize that much work is still needed. In some cases, relatively discrete gaps in the existing data can be filled by means of stand-alone studies. In other cases, however, a broader program of study will be required, particularly for projecting and managing social change. The committee was unable to provide cost estimates with confidence.

Conclusion 1: Although the demographic and economic effects of development-phase activities have received a good deal of competent attention, SESP does not appear to have dealt as successfully with assessing likely social and cultural impacts from development-phase activities or with identifying the steps that might be taken to manage or mitigate them. Partly as a result, not enough is yet known to permit the adequate assessment and management of the socioeconomic impacts that would be likely to result from development activities.

Recommendation 1: *Detailed attention should be paid to the range and scale of potential changes resulting from all phases of OCS activities, from MMS's prelease activities through the likely long-term impacts.*

Alternative: None recommended.

Conclusion 2: To date, studies generally have failed to reflect the fact that the human environment can be expected to change as soon as the potential for OCS-related activities is raised, often well before biological or physical disruptions begin. There is little evidence that systematic attention has been devoted to the fact that MMS can substantially ameliorate or exacerbate opportunity-threat impacts.

Recommendation 2: *The real (and often predictable and quantifiable) socioeconomic consequences of leasing and exploration-phase impacts need to be described and addressed.* MMS's influence over the magnitude and probable evaluation of these impacts will need to be assessed as well. The reasons for increased attention to this issue are pragmatic—i.e., related to successful prosecution of the leasing program—as well as scientific.

Alternative: None recommended.

Conclusion 3: MMS's socioeconomic studies have provided credible baseline assessments of the significance of subsistence activities and of the damage that could result if subsistence resources were damaged by a major oil spill. However, studies have included little attention to the alterations to subsistence activities that can take place even in the absence of oil spills and have not dealt systematically with the long-term or permanent socioeconomic changes (excluding oil spills, sudden shut-downs, or other acute or dramatic events) likely to result from development.   Given that the probability of major spills is much lower than that of less dramatic, but still significant, nonspill disruptions, greater attention needs to be devoted to the possibility that subsistence resources and activities can be disrupted even in the absence of spills and potentially even in the absence of production, as in the case of the controversy over the effects of drilling noise on the activities of the bowhead whale.

Recommendation 3a: *Commission social science studies that assess the impacts of OCS activities on subsistence and other significant sociocultural concerns likely to take place even in the absence of a spill.* Because the studies need to be credible to the affected populations and because the Iñupiat have a wealth of knowledge about the flora and fauna of the region, it is vital that greater effort be devoted to the cooperative development of studies or negotiated agreements with the North Slope Borough.   In addiiton, work should begin immediately, in close cooperation with the North Slope Borough, the state of Alaska, and affected citizens and their representatives, to address the scientific and pragmatic issues involved in dealing with the implications of the gradual, but cumulatively significant, adaptations that are likely to take place even in the absence of acute or dramatic disruptions.   Particular attention needs to be devoted to the assessing whether OCS oil and gas development might increase the possibility that prosperity and cultural vitality will continue well beyond the period of active exploration and development.   Otherwise, development could contribute to further intensification of the already serious potential for overadaptation and long-term socioeconomic stress.   This recommendation could not be considered to have been met if representatives of the North Slope Borough are merely offered the opportunity to provide data studies that remain fully within MMS control; instead, every effort needs to be made to carry out these studies with the full cooperation and collaboration of the North Slope Borough.

Alternative: None recommended.

Recommendation 3b: *Maintain an experienced and qualified staff of social scientists and secure adequate funding to carry out the necessary studies.*   Clearly, much of SESP's progress results from its historic ability to draw on a high-quality staff and on generally adequate funding. Unfortunately, in recent years, the MMS staff in Alaska has been cut dramatically. SESP had five professional staff through 1991; by July 1992, there were two (J. Imm, MMS, pers. commun., 1993). This level of staffing is insufficient to fulfill information needs. Furthermore, the recent loss of experience and institutional memory are significant concerns. At a minimum, SESP requires experienced and qualified staff members in the areas of anthropology, economics, and sociology.

Alternative: None recommended.

Recommendation 3c: *Seek to maintain and strengthen ties to the broader social science community.* Experience shows that there is no substitute for having at least a core of in-house staff members who are expert in relevant social science disciplines. Experience also shows that there are two important advantages in seeking to maintain and strengthen ties to the broader social science community. First, such ties can promote efficiency, permitting MMS to draw a far broader range of areas of specialized expertise than the agency could afford to support in-house. Second, they are vital in avoiding insularity and in maintaining the scientific credibility of study results.

Alternative: None recommended.

Recommendation 3d: *Ensure that quantification, analysis, and scientific conclusions are included in social and economics studies.* The technical personnel within MMS expressed to the committee during our meeting in Anchorage the belief that policy-making personnel discourage the drawing of conclusions about OCS-related impacts in socioeconomic studies, arguing that such decisions need to be left as a prerogative of policy makers. Just as it would not normally be considered acceptable scientifically to have a policy maker draw conclusions about the biological impacts of OCS development, it is scientifically inappropriate for technical social science extrapolations to be drawn by policy makers. Conclusions regarding effects on the human environment should come from scientific studies.

Alternative: None recommended.

Recommendation 3e: *Explicitly plan longitudinal, post-leasing studies, at least in areas where development appears likely to take place, at the outset.* This is a lesson that can be learned from MMS's experience in the Gulf of Mexico. Despite the clear differences across regions, at least some of the questions that cannot currently be answered about socioeconomic changes in Alaska are ones that could be addressed with greater confidence if a higher level of long-term, post-development research had been undertaken in the Gulf. Recent post-leasing studies there offer considerable promise for beginning to compensate at least partially for this problem, although some questions will not be answered with the same degree of confidence, or as efficiently, as might have been possible if the research had been carried out in a more timely manner.

Conclusion 4:   The studies focusing on mixed cash-subsistence economies provide an adequate description of baseline conditions. However, they fall short of providing a full analysis of potential impacts. For example, many studies describe subsistence activities and their cultural importance, but they do not attempt to analyze, quantify, and draw conclusions about the significance of possible changes in subsistence activities caused by specific scenarios for potential development.

Recommendation 4:   *To the extent feasible and appropriate, studies should quantify potential impacts to the subsistence sector of the economy using methods akin to those that quantify impacts to the cash sector.* The available MMS development scenarios should be used to identify areas lost to subsistence activities. This information could then be linked to information from MMS subsistence studies, which indicate where subsistence activities currently occur for each North Slope community. In this manner MMS could quantify the effects of potential land-use restrictions on subsistence activities for each community. Efforts also should be made to quantify other potential impacts to subsistence. This would require 2 person-months annually.

Alternative: None recommended.

Conclusion 5:   No studies attempt to determine whether or how alterations to the subsistence economy can be mitigated. The fact that cash obtained from employment is used to complement, rather than substitute for, subsistence harvests suggests that it will not be easy to mitigate losses in subsistence activity with monetary compensation.

Recommendation 5: *A thorough analysis of whether and how alternatives to subsistence activities can be mitigated is needed.* Studies also should address whether, in the face of a declining input of cash, Iñupiat communities will be able to revert to a primarily subsistence economy, or, if the communities will collapse, whether and how such an impact can be managed.

Alternative: MMS's decision-making documents should assume "worst-case" scenarios, i.e., that effects on subsistence may be unmitigable.

Conclusion 6: We concur with previous reviews that the rural Alaska model is well documented, replicable, and scientifically defensible. The economic impact model is adequate in terms of generic professional standards. However, the model is not well adapted for particular circumstances on the North Slope. Additionally, some concern has been expressed about whether the model adequately explains human migration patterns, particularly those of Alaska Natives.

Recommendation 6: *Economic growth models should be better adapted to the specific situation of the North Slope to measure cash flow to the local communities under various scenarios for development and for expenditure of resulting property revenues.*

Alternative: None recommended.

7 | MITIGATION AND REMEDIATION OF OIL AND GAS ACTIVITIES

INTRODUCTION

This chapter describes the potential effects of selected routine petroleum industry activities and accidental events in the marine arctic environment and their mitigation and remediation. In the case of accidents—the main concern is with oil spills—both the effects associated with placing equipment in arctic marine waters and the special difficulties in applying countermeasures to mitigate oil spills are discussed. To predict effects properly, one needs to know about the biological and human resources that might be affected and have some idea of the location and size of the hydrocarbon resource base. Prediction of the locations and severity of effects depends on the size, location, and distribution of the reservoirs to be developed. Loss or displacement of species harvested in traditional hunting and fishing areas must be considered in the evaluation of environmental effects. Seasonal extremes of temperature and light determine the periods of biological activity, and these factors also determine and limit the time available for control and cleanup of a spill.

There continues to be controversy, perhaps semantic in origin, about whether the arctic environment should be called "fragile." The popular news media and some special interest groups use the term to characterize the uniqueness of the Arctic on the evidence of long-lasting disturbance by human activities on land surfaces, although it is debatable if this rationale can be applied to industry effects in the marine environment. Some scientists with long experience in the Arctic (Dunbar, 1985) do not believe that the effects of pollution in the Arctic are greater than they are elsewhere. Dunbar makes the point that although there is less diversity among species in the Arctic than can be found in temperate and tropical regions, numbers within species tend to be larger and distributions extend over greater geographic areas, suggesting that even severe local damage could be repaired by immigration from adjacent areas. Also, it is known that the arctic environment can be highly variable within and among seasons, and it would not be an unreasonable assumption that arctic species are adapted to short-term harm and are able to compensate for population losses in one area or year by transmigration and adjustments in reproduction.

The danger lies in blaming human activity for a short-term loss in species caused, for example, by a heavy ice year. On the other hand, specific regions and populations can be extremely significant to Alaska Natives. In addition, long-term industrial presence in one area might have ecological consequences, especially on migrating species, by making this affected zone unsuitable for potentially obligatory habitat for a population. In that case, it would be essential to know the size of the "footprint" of the activity and its ecological spatial and temporal relevance to draw any meaningful conclusions about its potential to cause lasting harm.

AREAS AFFECTED BY ICE

Platform Design

The task of designing and constructing oil production platforms for regions such as the Chukchi and Beaufort seas should seem nearly impossible given the formidable environment that must be confronted, but it is not. It is generally believed that designing a permanent production facility is less of a challenge than is designing an exploration facility, because a production facility will be larger than an exploration platform. The object of exploration is to arrive on site, drill, and leave, encountering as little ice as possible. Careful timing of the drilling and strategies such as coming out of the hole or disconnecting the rig to avoid heavy ice are a common part of such operations. For production, periods of minimal ice would be chosen for moving a facility on site. However, once the structure is there, it would have to withstand the ice as it comes.

The oil industry already has considerable experience, for instance, with the platforms in Cook Inlet, which have successfully operated throughout the winter ice season for the past 27 years. Although the ice conditions in Cook Inlet are less severe than in the Arctic, the operations there nevertheless are instructive. There has been extensive exploratory drilling in both the Canadian and the Alaskan portions of the Beaufort Sea as well, where the systems used to protect against ice have ranged from artificially thickened ice pads that support winter-only operations in shallow water, to gravel islands, to gravel-filled caisson structures, to concrete structures. Several ancillary techniques also have been used to minimize the effect of ice forces on such structures. One is to construct berms that force grounding and deformation of the ice before it actually arrives at the offshore structure. Another is to spray water to build large, thick ice "bumpers" around drilling platforms. Although these techniques would prove useful in the development of designs for production platforms, they would be used only to minimize the general force that ice exerts on the platform, therefore minimizing structural fatigue. Platforms would have to be designed to deal with ice forces without the application of such techniques.

Designing offshore structures that can withstand the forces exerted by moving pack ice is hardly a new problem. Similar problems with designing icebreakers (Makarov, 1901; Lewis and Edwards, 1970; Weeks and Handy, 1970), in coping with ice forces on pilings and piers exerted by river and sea ice (Blenkarn, 1970; Afanas'ev et al., 1973; Neill, 1976; Nevel et al., 1977), and in evaluating ice-bearing capacity (Hertz, 1884; Wyman, 1950; Assur, 1961; Barthelemy, 1990) have been studied for years. In addition, since the discovery of oil at Prudhoe Bay, there has been a major industry program to advance knowledge about ice mechanics specifically related to the design of offshore structures.

Some sense of the extensive literature available on this subject can be gained by reading the work of Sanderson (1987, 1988), Cammaert and Muggeridge (1988), and Timco (1989). However, a few points need to be mentioned. There are many aspects of the general subject of ice forces on offshore structures that are less well understood than is desirable. For instance, there still is no generally agreed-on constitutive equation that describes the general mechanical behavior of sea ice. Also, the pressure-to-area curve for ice-and-structure interactions, although reasonably well documented at small scales (Sanderson, 1988), lacks an adequate theoretical basis to describe loads at failure as a function of the size of the failure area under consideration. High-quality biaxial and triaxial tests on sea ice are still quite rare. Tests that result in the full-scale failure of ice sheets are still limited, although a small number have been completed (Dykins, 1971; Vaudrey, 1977; Chen and Lee, 1986; Lee et al., 1986). Some observations have allowed estimates of the force levels reached when large, thick multiyear ice floes run into natural "towerlike" rock islands (Danielewicz and Blanchet, 1988; Montemurro and Sykes, 1989). An adequate explanation for the surprisingly low forces that were observed is still being sought. Models of the ridging process have not been adequately incorporated into procedures for estimating ice forces (Vivitrat and Kreider, 1981). However, even considering these limitations, there still are enough field and laboratory data and theory to assess threats to and design criteria of offshore production structures for the OCS areas of the Beaufort and Chukchi seas.

One reason that some believe that it is impossible to design production structures for the Beaufort and Chukchi seas is the presence of ice islands, which are produced by the breakup of portions of the Ellesmere Ice Shelf and are the tabular icebergs of the Arctic Ocean. They typically are 40-50 m thick with lateral dimensions that range from tens of meters to tens of kilometers (Jeffries, 1992). If a large ice island embedded in drifting pack ice were to run into an offshore production facility, the facility would probably be destroyed. The chances of this occurring are extremely small; large ice islands are rare. In addition, it is easy to identify ice islands by synthetic aperture radar (SAR) imagery (Jeffries and Sackinger, 1990). If an ice island were to drift near a production platform, the possibility of a collision would be known well in advance and proper precautions could be taken. Such rare events would, however, affect the potential economics of offshore operations and should be considered in estimating the economics of any new field.

It should be noted that research in ice mechanics continues. During the winters of 1992-1993 and 1994-1995, field programs supported by industry and by the Office of Naval Research (ONR) were and will be conducted. The ONR program also includes a laboratory component that is generating new test data under carefully controlled conditions. If successful, these programs should contribute new information about the scaling problem and add additional biaxial, triaxial, and full-ice-thickness tests. The committee does not believe that additional studies in ice mechanics supported by MMS are essential to the safe development of OCS oil in either the Chukchi or the Beaufort.

ICE GOUGING AND PIPELINE PLACEMENT

It takes several years to collect some of the information needed for design of safe offshore facilities. One example is the time required to collect the data necessary to make confident estimates of safe burial depths for offshore pipelines. A significant find that opens the way to development of an oil field must account for the way oil would be brought to shore. For example, if the oil is to be shipped by tankers loaded at the offshore field, feeder pipelines might be necessary to move the oil to a central loading point. In the Beaufort Sea and at most locations in the Chukchi Sea, feeder pipelines would probably transport the oil near to the center of the field, from which it would be fed into a main pipeline for transport to the coast and ultimately to the Trans-Alaska Pipeline System. In the Chukchi Sea, it is easy to conceive of production sites requiring hundreds of kilometers of subsea pipelines that transect regions where the seafloor is extensively gouged by ice. In the Beaufort Sea, the area affected by gouging is smaller, because the shelf is narrower, but the problem is no less severe; gouging is generally more intense in the Beaufort Sea. Gouging is not a problem in the deeper Navarin Basin.

Engineers designing such subsea pipelines need to be able to answer several questions, including the following:

• How deep should the line be buried along each section of a proposed pipeline route to ensure that the risk of its being damaged by a moving ice mass is acceptable?

• Is simple burial adequate for absolute safety in high-risk areas or is it necessary also to armor the pipeline to reduce the possibility of an under-ice spill from a ruptured line?

• Are there regions where gouges do not occur and where pipelines can be placed on the seafloor with minimal burial or none at all? If so, where are these regions?

Given the cost of deep burial and the short operating season for pipeline burial equipment such as subsea plows or cutter-suction dredges, the cost of pipeline installation could have an appreciable effect both on when a given field could be brought on line and on the overall economics of the field. In an extreme situation, pipeline costs could rule out production at some sites or, more likely, could delay the decision-making process while missing data thought necessary for safe design were collected.

State of Knowledge

As the ice pack moves over the shallower waters of the continental shelf, the keels of pressure ridges and the lower surfaces of ice islands plow up the seafloor has been known as a scientific curiousity since the 1920s. Once offshore oil development on the margins of the Arctic Ocean became a possibility, a thorough understanding of the process became important. During the past 20 years, gouging has received considerable attention. For instance, a recent bibliography (Goodwin et al., 1985) lists 379 publications with some 101 papers dealing with the Beaufort Sea, 23 that discuss the Chukchi and Bering seas, 15 that deal with theory and modeling, and 29 that deal with protection. (There also are many published studies about gouges produced by true icebergs off the east coast of Canada. Most of this information is not particularly applicable to arctic regions.)

It is well established that gouge depths can be described by a simple negative exponential distribution: Small gouges are frequent; deep gouges are rare. The spacing between gouges also shows a similar distribution, suggesting that gouging can be approximated as a random process. Both the frequency and the intensity of gouging are clearly related to water depth. In the U.S. portion of the Beaufort Sea, most deep gouges occur in water of 30-40 m deep. The largest gouges observed on the U.S. portion of the Beaufort shelf are about 5.5 m deep (Barnes et al., 1984). Larger gouges of up to 6 m deep have been observed in the Canadian portion of the Beaufort Sea, probably because of differences in the consolidation of the subsea sediments. Closer to the Mackenzie Delta, gouge densities are extremely variable; there can be as many as 500 gouges/km^2. Track samples indicating more than 100/km^2 are not rare. In the Chukchi Sea, the deepest gouges occur at water depths of 35-50 m and are as much as 4.5 m deep. Most gouges are oriented roughly parallel to the coast, running generally east-west in the Beaufort Sea and northeast-southwest in the Chukchi Sea. In both the Beaufort and the Chukchi seas, where there are offshore shoals, the largest number of gouges occur on the seaward flanks of the shoals (Reimnitz and Kempema, 1984). In general, gouging is rare in water that is more than 50 m deep, and it does not appear to occur in water deeper than 60 m. Unfortunately, the amount of data bearing on these limiting values is small.

Improving the State of Knowledge

It would be desirable to have a better estimate of the depth at which gouging ceases to be a problem in the Chukchi and Beaufort seas. In the recent geological past, sea level was roughly 100 m lower than it is now. Therefore, gouges in deeper waters may be relics rather than indicative of current activity. This is particularly true if gouges formed in deeper water survive for long periods. Because the Chukchi and Beaufort shelves are shallow and slopes are low, small changes in the maximum water depth where active gouging occurs result in large changes in the areas where gouging must be considered a problem.

Even more important is the limited availability of data that provide direct determination of gouging rates as a function of water depth and location. It is impossible to know, for example, in a region with 100 gouges/km^2, whether they were formed during the past year, during the past 10 years, or during the past 1,000 years. There is no absolute way to date gouges, despite the importance of the information to the decision-making process (Weeks et al., 1984; Wang, 1990). Simulations suggest that most gouges are filled within a few years (Weeks et al., 1985). Field observations also suggest significant sediment movement along the arctic OCS (Barnes and Reimnitz, 1974; Barnes, 1979). Whether these conclusions apply to water depths greater than 45 m is not known.

Measurement Program

The development of a good data base on gouging rates and the infilling process requires yearly replicate measurements of gouging patterns in selected regions of the Alaskan OCS. Although some data exist (Barnes et al., 1978; Rearic et al., 1981), expansion of this data base would contribute significantly to safe pipeline design. We stress that this information cannot be quickly obtained even if major financial resources are devoted. On the other hand, by carrying on a low-level, continuing program, excellent data can be obtained for reasonable cost. The longer the time series, the better will be the estimates of the gouging rates and initial gouge depths.

This recommended program will not replace the site-specific measurements that are required once a field large enough to warrant development is found. Nevertheless, general information would allow designers to view the site-specific measurements, which undoubtedly would be of much shorter duration, in a broader context. Given that ice-induced ruptures of subsea pipelines should be considered a probable cause of below-ice oil spills and that current practice assumes that the granting of offshore leases will be followed by permission to develop if adequate petroleum resources are found, the more timely and safe development that should result from of these proposed measurements would be well worth the additional cost.

The importance of initiating this program promptly is supported by the following observations. ARCO announced in October 1992 the discovery of oil (3,400 barrels/day) at its Kuvlum No. 1 drilling site. The well is located in 36 m of water on the Beaufort shelf, roughly halfway between Prudhoe Bay and Kaktovik (Barter Island). The drill ship had to discontinue operations several times and move off the site to avoid collision with an extremely large, thick floe drifting nearby. (This procedure is normal, effective, and necessary; drill ships are not designed to survive significant ice forces.) During the time the floe was in view of the drilling platform, there was a period when the floe was obviously grounded, presumably resulting in one or more gouges. The water depth at the grounding location was 30 m. The Kuvlum prospect is believed to be part of a field that could produce more than 1 billion barrels of oil. To date in the Beaufort OCS, 8 of 21 wells drilled have exhibited significant quantities of oil, so we believe that it is desirable to know as much as possible about gouging over both the Beaufort and Chukchi OCS.

ARCTIC OIL SPILLS:
INCIDENCE AND RESPONSE

Recent reviews of the quantity and incidence of oil discharges into the world's oceans (GESAMP, 1993) show that input of oil from anthropogenic sources has decreased during the past decade. The quantity of oil entering the ocean from tanker spills has declined to only about 20% of input. (NRC, 1991c). Although marine transportation accidents account for many of the largest single accidents, their frequency has declined. Figures 7-1 and 7-2 demonstrate this trend, particularly in the United States in recent years. Transportation accidents presumably will continue to decline as stronger regulatory controls are implemented.

The Canada Oil and Gas Lands Administration (COGLA, 1985b) published a review of the question of blowout frequency associated with petroleum drilling, based on worldwide records for 1955-1980. The study

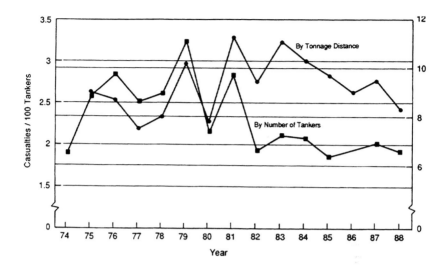

FIGURE 7-1   Worldwide rates of serious casualties to oil/chemical tankers, 1974-1988 (6,000 gross tons and above). Source: NRC, 1991c.

determined that the blowout frequency associated with an oil loss was 1 in 3,055 wells drilled; the incidence of blowouts associated with a "significant" oil loss—more than 50,000 barrels—was more remote, 1 in 7,325. The accumulated incidence of blowouts at exploratory wells was about three times as high as that for development wells.

The frequency of pipeline ruptures worldwide has not decreased in recent years, resulting in a greater proportional contribution to the total amount spilled (Figure 7-3) (OSIR, 1993).

Spill Response

The response to an oil spill in arctic conditions is complicated by three main factors: the common presence of ice, the distance from spill cleanup logistic support, and long periods of darkness and cold. Although the response to a spill in open water can rely on standard techniques such as mechanical containment and recovery, ice-infested waters present special

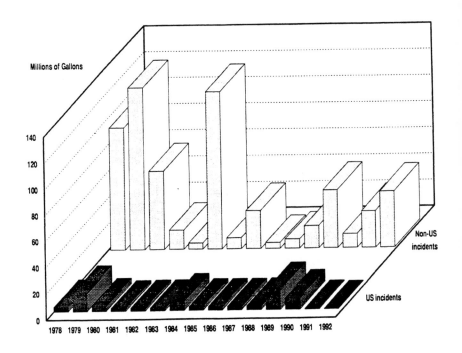

FIGURE 7-2   Annual amounts spilled in vessel incidents, 1978-1992
(includes tankers, barges, and other vessels). Source: OSIR, 1993.
Reprinted with permission from *Oil Spill Intelligence Report Newsletter;*
copyright 1993, Cutter Information Corp., Arlington, Mass.

problems and the effectiveness of most containment and recovery systems
is compromised. New technology is slow in being realized, although burning oil on ice and in ice-enclosed water surfaces appear to be the most
useful countermeasure. An ice corral can contain oil so that countermeasures can be applied, but it also can prevent access. Mechanical containment and recovery have limited ability to handle solids (such as broken ice).
Limits on remote-sensing capacity and difficulty in modeling the fate of oil
caused by inadequate detailed knowledge of under-ice currents are additional problems. Finally, bioremediation attempts are likely to be slowed
by the low water and shoreline temperatures. When oil is spilled, the first

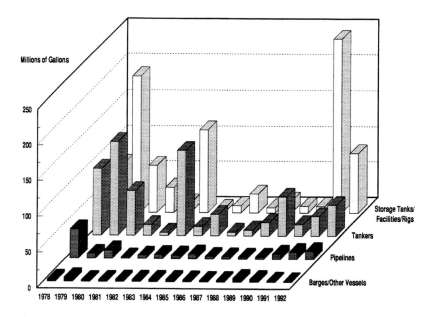

FIGURE 7-3   Annual amounts spilled by source type, 1978-1992. Source:
OSIR, 1993. Reprinted with permission from *Oil Spill Intelligence
Report Newsletter;* copyright 1993, Cutter Information Corp., Arlington,
Mass.

priority in the Arctic, as elsewhere, should be to stop the source of the
release and to contain the spill to as small an area as possible (Mackay,
1985).

Spills in Ice and in Very Cold Water

The occurrence and fate of arctic oil spills have received considerable
attention in the technical literature since oil was discovered at Prudhoe Bay
and in the Canadian Arctic. The following books and papers offer an in-

troduction to this subject: Malins (1977), Clark and Finley (1977), Thomas (1983), Engelhardt (1985a,b), Mackay (1985), Payne et al. (1987, 1989), Baker et al. (1990), annual Arctic Marine Oilspill Program symposia, and biennial oil spill conferences.

For arctic oil spills, two possibilities should be examined: first, that both the spill and the cleanup occur under ice-free conditions, and ice is not present during the interim period; second, that sea ice is present at some time during the life of the spill.

Spills in Ice-Free Water

When both the spill and the cleanup occur in ice-free water, more information and response capability are available because the expected environmental conditions will not be much different from those that have been encountered at marine spills in more temperate regions. The primary differences here are well understood and can be attributed to the fact that the sea temperature is never far from the freezing temperature of sea water ($\sim$-1.8° C). Therefore, after the spilled oil cools to ambient temperatures (-1 to +5° C) the oil will be more viscous than at higher temperatures. In addition, any chemical reactions that lead to the degradation of oil will be slower. Typically, degradation at 5° C is 20 to 50 times slower than at 25° C (NFAC, 1978); for instance, the loss of volatile organic compounds from the oil will take longer (Kirstein and Redding, 1987; Payne et al., 1987), which can lengthen the period during which the oil can be burned successfully. At the same time, the lower temperatures make ignition more difficult and burning proceeds at a slower rate. Burning has been used with some success in test spills in the Canadian and Norwegian Arctic, but additional tests are desirable. More important is that even though cleanup techniques are more efficient in temperate regions than they would be in the Arctic, recovery of oil from very large spills even in temperate waters is unlikely to exceed 20%.

Spills in Sea Ice

There are several scenarios for spills in regions with sea ice. For instance, the spill could occur in open water and new ice could develop before cleanup begins, or an ice cover could already be present at the time of the spill. Also, ice conditions might be such that cleanup could be accomplished completely in the ice or only after the ice had melted or, more likely, by a bit of both. Finally, a spill could happen near fast ice or in mobile pack ice—two vastly different situations.

Spills in Newly Forming Sea Ice

The simplest example in which the spill occurs in open water and subsequently becomes involved in newly forming sea ice occurs during the initial fall freeze-up when a thin first-year ice cover starts to develop. Good examples of regions where this is possible are the Bering and southern Chukchi seas; both are commonly completely ice-free at the end of the summer. However, similar processes can occur in polynya areas or in oil-covered leads even in regions of heavy multiyear pack ice. In such situations, the presence of the oil will damp surface waves and reduce turbulence in the upper part of the water column, thereby reducing conditions that would favor the formation of frazil ice, the fine-grained, granular ice that forms in supercooled, turbulent water. Once the oil becomes cold enough, sea ice starts to form beneath the slick, and most of the oil will be trapped either on the upper ice surface or in the top few centimeters of the ice (Payne et al., 1987). Any rafting, which is common during this growth stage, will presumably smear oil over both the upper and the lower ice surfaces. In the case of pressure ridging, the oil will be incorporated completely throughout the ridge. It would then move with the ice and could be tracked along with the ice. Cleanup during this time of the year (largely during fall freeze-up) would be difficult at best. Although the ice is thick enough to limit shipping to vessels specifically designed for ice operations, it is still too thin to support heavy equipment. Also, such ice is frequently very mobile and commonly experiences lead formation and ridge building. During freeze-up the weather is commonly marginal and daylight is limited, so the problem of deploying and guaranteeing the safety of cleanup crews would become a major consideration.

Spills in Fast Ice

If a spill occurs in fast ice, some advantages can be had if the ice is thick enough to support cleanup equipment. For instance, if the spill is on top of the ice, the oil will be restricted to a smaller area. However, the oil will seep into the surface snow cover, and if the spill is of an appreciable size, the task of removing large volumes of oil-saturated snow is formidable. Also, in surface spills, some of the oil would invariably become trapped either within or below the ice.

In each case, locating and cleaning up the fraction of the spill either within or below the ice would be difficult. In all probability, this oil would not be recovered until the ice cover melts, at which time open-water cleanup procedures could be used. By this time, the fast-ice region will have become a region of pack ice. If the spill occurs below a region of fast ice, the ice also tends to restrict the movement of the oil, because the underside of undeformed fast ice is irregular due to local variations in ice growth rate caused by variations in the thickness and properties of the surface snow (Kovacs, 1977; Barnes, 1979). A few studies suggest that the bottom roughness of typical fast ice near the margin of the Arctic Ocean is adequate to immobilize significant quantities of oil. Estimates of the volume that would be trapped by the rough underside of the ice and therefore that would drift with the ice are 10,000 to 35,000 m^3/km^2. The containment potential for multiyear ice is generally believed to be much larger. For instance, Kovacs (1977) estimates a value of slightly under 300,000 m^3/km^2 for multiyear ice with a mean draft of 4.3 m. The exact geometry and leakiness of these under-ice pools has not been documented.

If the oil remains in place for an appreciable period during the ice-growth period, subsequent ice growth below the oil layer will result in the incorporation of the oil into the ice sheet. The problem is then with locating and removing the oil. We are not aware of a proven remote-sensing method for locating such oil, although it is possible that either surface-based radar systems (Kovacs and Morey, 1986) or helicopter-borne electromagnetic systems (Kovacs and Holladay, 1990) will prove useful. Drilling and coring would be effective, but these procedures are not efficient.

Once the oil is located some of it can be removed by down-hole pumping techniques. There is no simple way to remove oil that is entrapped within ice, so final cleanup will be possible only after the ice has melted and open-water or on-ice procedures can be applied. Encapsulation of oil by subsequent ice growth is only one method by which oil can become trapped in the ice. If the oil is of a sufficiently low viscosity it can move upward into the ice through the ubiquitous brine drainage channels that occur in natural sea ice (Lake and Lewis, 1970). In 2-m-thick sea ice, the average spacing between these brine channels is about 15 cm (Lewis, 1976). Brine drainage channels are particularly important in the spring because they can enlarge to features with a diameter of 5 cm or more, significantly facilitating the entrance of the oil into the ice. Unfortunately, although brine drainage tubes have been studied to explore their use in the desalination of sea ice, their role as a pathway to the incorporation of oil into sea ice has received little attention. Oil can also accumulate in the much larger diameter seal-breathing holes, presumably filling them up to at least the thickness of the ice (the actual position and height of the oil in the holes will depend on the volume of contiguous oil under the ice surface, the density differential, and the size of the opening). This clearly would threaten seals if the spill were large.

By summer, oil would be on top of the ice even if the initial spill were initially completely entrapped beneath the ice (Clark and Finley, 1982). As increasing amounts of shortwave radiation arrived both at the ice surface and at locations within the ice, this would warm the oil relative to the ice because of oil's much lower albedo—the fraction of incident electromagnetic radiation reflected by the surface. The shortwave radiation also would speed the melting of the ice, which would speed up the time when open-water cleanup techniques could be applied. The Office of Technology Assessment (OTA) estimates that mechanical recovery has been approximately 15% efficient under good conditions. Even though major advances are being made for spill response in open waters as a result of the requirements under the Oil Pollution Act of 1990, it is unlikely that the capability for control and recovery of spilled oil in ice-infested waters will be easily improved

Spills in Pack Ice

The most difficult situation is when the spill occurs in a region of heavy, mobile pack ice of varied thickness—the situation most commonly encountered in the Beaufort and Chukchi seas, where multiyear ice is invariably present. Also associated with this is the ubiquitous occurrence of leads and pressure ridges. Of particular concern is the nearshore lead system—sometimes referred to as the flaw lead or leads—located just seaward of the outermost fast-ice edge, which during the spring serves as a seabird foraging area and marine mammal migration route. The polynya that commonly occurs along portions of the Chukchi coast under conditions of offshore winds is important because it is an area of significant seabird and marine mammal activity.

Cleanup of all but the smallest spills in pack ice presents a major problem, even though the ice itself tends to immobilize all or a portion of the oil, depending on the size and the nature of the spill. However, this is a small consolation: The ice in which the oil is spilled is, in most cases, drifting to the west at speeds of 2 to 3 km/day. During storms, drift rates on the order of 2 m/s (4 knots) have been observed by radar at Barrow, Alaska (Shapiro and Barnes, 1991). This could make the logistical aspects of a cleanup extremely difficult. The fate of the oil in the ice is similar to what might be anticipated from our previous discussion of oil in fast-melting ice, with the addition of the complications added by extensive lead formation and ridging. Leads occur not only at the fast-ice edge, they can be found anywhere within the pack. Any oil in the water that is not trapped by the bottom roughness of the ice has a good chance of ending up in an area where the ice is thinnest; in fact where there is little or no ice at all—in a lead. Once in a lead, the oil is effectively trapped, although it can readily move laterally in the lead. The oil can, in principle, form a thick layer because during the winter leads typically comprise only a small portion of the area of pack ice. Once trapped in a lead, the oil can be moved by the wind to the downwind side of the lead. There, it can be incorporated into the accumulating frazil or form a layer on top of newly forming ice as described earlier.

As the thinner ice classes characteristically found in leads are also the ice most commonly found in pressure ridges, it is to be expected that numerous ridges will have oiled ice distributed throughout their complete thickness. It is also likely that the keels of pressure ridges, which can go as much as 50 m into the water, will act as skimmers to any oil present in the water column that is not effectively fixed to the ice. This will lead to an accumulation of oil on the upstream side of the keel.

Another difficulty is caused by the presence of the thicker (2.5-5.0 m) undeformed multiyear ice, which can constitute a major percentage of the ice cover. The additional thickness increases the difficulty of locating oil beneath the ice. Also, in many cases the ice floes are neither large enough nor flat enough to make the use of heavy equipment practical, even though they might be of sufficient thickness to support the equipment.

In short, the presence of ice greatly complicates the already difficult problem of cleaning up an oil spill in the ocean. In fact, it makes cleanup so difficult that in many cases final cleanup will be possible only after the region of ice containing the oil has melted and open-water procedures can be applied. It is therefore important to be able to track the affected region of ice until it melts.

Tracking Spills in Ice

The direct detection and tracking of oil in and under ice is not feasible with any certainty at this time. There are several techniques available for tracking ice. Perhaps the simplest is to place one or more global positioning system (GPS) buoys on the affected ice. This would provide a continuous record of the position of the ice until final cleanup. If it is not possible to do this, the movement of the ice could be tracked by SAR imaging. Adequate tracking should be available after 1995, when the Radarsat satellite is scheduled to be launched. It will provide the wide-area coverage (500-km) and high resolution modes that are lacking in the current generation of SAR satellites. Open-ocean oil slicks can be identified by SAR, which therefore could be used to both track the ice and to identify the final open-ocean oil slick. Undoubtedly, both GPS and SAR would be used. There is also some promise for a sensor that can be used to detect oil under ice; for example, a pulsed ultraviolet fluorescence sensor that could detect oil through ice of 0.5-1 m thick (Moir and Yetman, 1993).

The destination of an oil spill originating in the Chukchi-Beaufort region is not predictable (Thomas, 1983; Colony, 1985; Pritchard and Hanzlick, 1987). Depending on the location and timing of the spill, the oil could exit the Arctic Ocean via the Bering Strait and arrive at the ice edge somewhere in the Bering Sea. However, it is more likely to drift to the west past Wrangel Island (Colony, 1985), where it could either hit the Siberian coast or become involved in the drift of the pack in the Transpolar Drift Stream. This route would take the oiled ice over the North Pole toward Fram Strait between Greenland and Svalbard (an archipelago of Norway), allowing the oil to exit the pack some 2-4 years later into the North Atlantic off the east coast of Greenland (Thorndike, 1986). It is also possible that the oiled ice could become trapped in the Beaufort Gyre for 1-2 rotations of 5-10 years each before ultimately joining the Transpolar Drift Stream. When production areas are identified in the Chukchi and Beaufort seas, additional studies that use state-of-the-art ice-dynamics models should be completed and compared with data sets collected via drift buoys to validate the forecasts of ice trajectories for these areas.

EFFECTIVENESS OF RESPONSE MEASURES

It is important to evaluate spill countermeasures in an ecological context. Ever since the Torrey Canyon (1967) and Santa Barbara (1969) oil spills, there have been conflicts during spill cleanup over the desire to remove visible oil so that the environment looks "clean" and the desire to minimize the spill's environmental impact and promote natural recovery, which can mean using control methods in one area to prevent oil from entering another or leaving some oil on shorelines or in sensitive habitats when further cleanup would increase the damage caused by a spill (Lindstedt-Siva, 1979a,b).

Much research has been done on the effects of spilled oil and of various spill cleanup methods. A review paper prepared for a 1982 workshop of oil spill specialists cited approximately 1,000 references (Tetra Tech, 1985). The report was used as a basis for discussions, and the recommendations developed were later published as a manual (API, 1985) that identified habitat types that are particularly sensitive to damage from spills and techniques to protect them. The manual also evaluated the relative damage caused by various shoreline cleanup methods compared with the damage caused by the actual spill. Steam cleaning and other high-impact methods generally were not recommended. Low-pressure flushing and other low-effect techniques were recommended. The manual also promoted the rationale that it is often better to leave some oil in the environment than it is to use extreme measures to remove it.

Response Technologies

A work group on oil spill cleanup (Lindstedt-Siva, 1984) divided methods into several generic types, among them natural cleansing, mechanical removal, use of sorbents, use of chemical dispersants, use of sinking agents, and burning. Four general categories of environmental damage that can result from these methods also were identified. Table 7-1

TABLE 7-1   Potential Impacts on Various Oil Spill Cleanup Techniques in Different Habitats

Clean-up Methods	Fine-Grained Beaches	Coarse-Grained Beaches	Mixed-Grain Beaches	Rocky Shores	Mud Tidal Flats	Sand Tidal Flats	Salt Marshes	Mangroves	Mammal & Bird Rookeries	Sea Grasses	Coral Reefs	Kelp Beds	Rock Reefs	Oyster Beds	Flowing Water, Non-Organic Shorelines	Flowing Water, Vegetated Shorelines	Lakes, Non-Organic Shorelines	Lakes, Vegetated Shorelines
1. Natural Cleansing	+1-4	+1-4	+1-4	+1-4	+1-4	+1-4	+1-4	+1-4	+1-4	+1-4	+1-4	+1-4	+1-4	+1-4	+1-4	+1-4	+1-4	+1-4
2. Mechanical Removal																		
a. Containment & Recovery									+1	+1,2	+	+	+	+	+1	+1	+1	+
b. Mechanized Equipment	+1-4	+1-4	+1-4		+1-4	+1-4	+1-4			+1,2,3					+1		+	
c. Substrate Disturbance	+1-4	+1-4	+1-4				+1-4											
d. Beach Cleaners	+		+															
e. Vacuum Pumping	+		+						+1,2						+	+	+	+
f. High Pressure Flushing				+2						+1,2					+	+		
g. Low Pressure Flushing	+1	+1	+1	+	+1	+1	+	+	+						+	+	+	+
h. Steam Cleaning				+2,3,4						+1,2						+1,2		+1,2
3. Manual Removal																		
a. Collection	+1,2	+1,2		+2	+1,2	+1,2	+1,2	+	+2									
b. Vegetation Cropping				+2				+1,2	+2	+1,2								
4. Sorbents	+	+		+	+3,4	+3,4	+3,4	+	+						+	+	+	+
5. Chemical Dispersants	+? 3,4	+? 3,4	+? 3,4	+3,4			+3,4				+3	+3	+3	+3	+3	+3		+3
6. Sinking Agents										+1,2,3	+1,2,3	+1,2,3	+1,2,3	+1,2,3				
7. Burning							+1,2,3,4											

+ - Technique has been used or considered
Potential Cleanup Impacts
1 - Potential for substrate or physical habitat disruption.
2 - Potential for loss or disruption of community structure.
3 - Potential for increase in likelihood of oil uptake potential.
4 - Potential for increase in the likelihood of oil re-entry to the environment.   *modified from (8).

summarizes the areas of potential harm in the four categories by various cleanup techniques in 18 habitat types.

• Substrate or habitat disruption. Some cleanup methods disrupt the integrity or change the quality of a substrate, and, consequently, associated organisms. One example is the use of heavy equipment on some beaches.

• Loss or disruption of community structure. Cleanup techniques, such as steam cleaning or high-pressure flushing of a rocky intertidal habitat, can kill organisms that survive the spill itself.
• Increased likelihood of oil exposure or uptake. This occurs when the cleanup technique, such as high-pressure, hot-water washing, exposes organisms to additional oil because of increased penetration of oil into sediments.
• Increased oil reentry to the environment. This can occur with techniques that reduce the viscosity of the oil (hot-water treatment), disturb an oiled substrate to remobilize oil (flushing without proper recovery of free oil), or incorporate oil (steam cleaning) into sediments where, as erosion occurs, it will be reintroduced into the environment.

Mechanical Removal

Mechanical removal methods include the use of special equipment developed to contain and recover oil on water, as well as conventional construction or other motorized equipment (vacuum trucks, front-end loaders) adapted for oil spill cleanup.

Mechanical removal can remove organisms as well as sediments from shorelines. There is the potential for damage to surrounding habitats from operation of heavy equipment or from trampling by work crews. Many Alaskan beaches are so remote as to make mechanical cleanup infeasible.

Dispersants

There has been much research, including several experimental spills (NRC, 1989b), on the effectiveness and the effects of dispersant use. Generally, the tradeoff is between the relatively short-term effects of a pulse of dispersed oil introduced into the water column versus the long-term effects of oil that spreads out over broad areas, strands on shorelines, and contaminates wildlife. When applied by aircraft, dispersants, which are less toxic than oil when applied at sufficiently low concentrations, can cover a large slick relatively quickly and are effective enough to affect the environmental outcome of the spill. One National Research Council panel recommended that dispersants be considered equally with other response measures as a means to increase the overall effectiveness of spill response (NRC, 1989b). Sergy (1985), in a review of the results of the Baffin Island Oil Spill project—a controlled set of experimental spills in the Canadian Arctic—concluded that there are no compelling reasons that dispersants should not be used in the Arctic when their use would result in benefits relative to other options. The American Society for Testing and Materials (ASTM, 1986) has developed guidelines for dispersant use in the Arctic.

Burning

In-situ burning of oil on ice was first developed as a response method for the Arctic for oil on the ice surface. There has been a large continuing program in recent years to evaluate burning as a major on-water response method. Results indicate this method is effective (Allen, 1988). It may be the most effective means to deal with oil on ice, in ice leads, and in snow (Sveum and Bech, 1991; Brown and Goodman, 1986). For example, MMS has recently supported useful, multiple-sponsored studies of in-situ burning as a spill response (Evans, 1991). Test burns of oil in ice were conducted in Spitsbergen, Norway, in 1992, and were planned again for May 1993. A large complex in-situ burn study on open cold waters was carried out in August 1993 off Newfoundland.

Bioremediation

Bioremediation—enhancement of natural biodegradation processes—shows promise as a shoreline post-spill cleanup method, but further research is needed to assess its effectiveness under arctic conditions and to make products suitable for the Arctic. The field of bioremediation of oil on shorelines is currently under active research and development in both public and private sectors.

Shoreline Protection and Cleanup

The implications of spill responses in various habitat types were discussed in the 1970s (Westree, 1977; Lindstedt-Siva, 1977, 1979a, b). The major contributions of this work were to identify habitats that were particularly vulnerable to the effects of spilled oil and to assess the effects of various cleanup techniques, as well as the effects of oil.

A 1982 workshop, which included international oil spill experts from government, industry, consulting firms, and universities, produced a manual, "Oil Spill Response-Options for Minimizing Adverse Ecological Effects" (API, 1985). This publication evaluated response techniques in terms of their ecological effects as well as their effectiveness, and presented recommendations. It also identified environmentally sensitive habitats recommended for high-priority protection during a spill. The manual contains a section on arctic habitats.

Ecology-based response planning includes an analysis of a particular segment of coastline, identification of specific environmentally sensitive habitats, and development of plans to protect them—to prevent or reduce the amount of oil entering them if a spill should occur (Hum, 1977; Lindstedt-Siva, 1977; Michel et al., 1978; Pavia et al., 1982). Ecology-based response planning's goal is to minimize the environmental damage caused by spills by protecting those areas most vulnerable to longer-term effects. Much of the arctic Alaskan coastline has been analyzed by the state of Alaska and the oil industry (Alaska Department of Fish and Game, 1977; Alaska Clean Seas, 1983).

Experimental Spills

One of the best ways to study the fate, behavior, and effects of oil spills, assess the effectiveness and effects of countermeasures, and improve the capability to respond effectively is to conduct experimental spills (controlled releases). In recent years it has become increasingly difficult to obtain permits for such research in US waters. This approach has been recommended for basic research, evaluation of response techniques and training (Interagency Coordinating Committee, 1992; NRC, 1993b; Owens et al., 1993). There are many advantages of conducting experimental spills over using so-called "spills of opportunity." First, there is opportunity to plan the experiment to fit the site. This includes collection of prespill data (something usually not possible at spills of opportunity) and control of the dosage of oil and the exposure of ecological communities. This allows researchers to distinguish more subtle changes than would be possible during accidental spill events and allows verification of laboratory observations.

One of the most important advantages of well planned experimental spills is that teams of scientists and the required logistical support can be mobilized and dedicated to the project. It is difficult at best to conduct science during the emergency phase of the response to accidental spills. During a complex response operation, there are many priorities (e.g., control of the spill at the source of the release, vessel stabilization, offshore recovery operations, protection of environmentally sensitive areas) and limited resources (e.g., aircraft, vessels). Science, understandably, will not be a priority during this phase. Nor is it always possible to mobilize the scientific personnel and equipment needed rapidly. This is certainly an important consideration in the Arctic.

Experimental spills also are needed to evaluate spill- response technologies. For example, some testing of mechanical recovery equipment and chemical dispersants can be conducted in test tanks and in the laboratory. However, for full evaluation of these methods at appropriate scales, occasional tests must be conducted in the field. For reasons described above, it is usually not possible to conduct these evaluations adequately during a spill emergency, nor should such evaluation be a condition of using these methods.

Finally, experimental spills are needed to train people who must respond to accidental spills. If response crews are never allowed to practice on real oil, they cannot realistically be expected to perform well during an actual spill emergency.

RESTORATION

During the past decade, restoration ecology has become recognized as an important and integral component of a national interest in conservation (NRC, 1992c). Attempts to restore ecosystems follow environmentally damaging events or a process of degradation; they assume recognition of the effects and a desire to accelerate recovery. Effective restoration requires knowledge of biological detail and the outcome of interactions between species. It is a test of our knowledge of the important interactions that determine the form and functional relationships within communities. It is a challenging task, and often we do not know enough to be successful (NRC, 1992c).

On the North Slope, efforts to vegetate disturbed areas with native grasses used by foraging caribou provide an example of research designed to anticipate a needed solution rather than merely to document injury. Restoration or maintenance of valued resources could also require manipulations of animal populations that expand abnormally around areas of development.

Given the limitations in our ability to restore damaged ecosystems, and the difficulty in identifying aspects of marine ecosystems that might be amenable to restoration, what should MMS do? First, it should investigate available information on the restoration of arctic ecosystems and identify research needed to protect and restore, where possible, environments at risk of damage from OCS oil and gas activities. Such research should include the identification of those habitats and ecosystems that are unlikely to be amenable to restoration efforts; indeed, it seems safe to say that MMS should focus on nearshore and terrestrial ecosystems rather than on deep-water habitats and pelagic ecosystems. Second, MMS should take advantage of restoration efforts of industry on the North Slope and evaluate its experiences. Finally, although this study's scope includes only Alaska, the committee recommends that MMS consider restoration as part of its national framework and use experience gained in any region to inform its efforts in other regions.

MITIGATING SOCIOECONOMIC EFFECTS

Although MMS-sponsored studies have often been sensitive and accurate in discussing the importance of subsistence activities, they have generally been weak in identifying the steps, if any, that could be taken to mitigate negative impacts on subsistence, whether those impacts are spill-related or take place in the absence of a spill. In general, the studies tend not to go beyond broad or overall assessments, outlining instead the significance of subsistence to Iñupiat ways of life, and pointing out that if subsistence resources were ever affected by a spill, the consequences could be almost overwhelmingly negative. The fact that cash obtained from employment is used to complement, rather than substitute for, subsistence harvests suggests that it will not be easy to mitigate losses in subsistence activity with monetary compensation.

Similarly, many of the studies describe the "boom" aspect of the oil-related cash economy, and they project a coming "bust." Some studies note that OCS development may forestall, but not prevent, the forthcoming decline. However, none of the studies systematically addresses the potential for mitigating the impact of the coming "bust" on the communities, nor do they discuss potential options for dealing with the more gradual problem of long-term overadaptation or offer options for managing the inevitable impacts. Studies also need to be conducted that address whether the Iñupiat community will be able to revert to a primarily subsistence economy with a declining input of cash, or whether the communities will to some degree collapse. They also need to deal with the questions of whether and how such an effect can be managed. For example, Barrow has accumulated a large infrastructure and associated maintenance costs, as well as a substantial debt, a factor that points to the need for further consideration of pragmatic impact-mitigation options. In order to manage impacts to the human environment, it is critical to determine how to deal with debt payments as oil revenues begin to decline and ultimately cease.

Significant numbers of Alaska Natives have been employed by the borough—often further contributing to local betterment through the construction of housing, educational, and sanitation facilities; through improvements in health and social services; and through activities that help document, strengthen, and preserve traditional cultures. All of these activities, and more, have been facilitated in part by oil revenues from Prudhoe Bay. Depending on the ways in which any potential OCS development would be managed, it could contribute either to a worsening of the eventual "bust," with the cessation of activities in the region, or to the effective mitigation of those problems.

As another example, it is possible that oil development in OCS waters could help to sustain oil-related activity in the area by extending the useful life of the Trans-Alaska Pipeline Systems. Unlike the case for the Prudhoe Bay and onshore fields, however, developments in federal waters would not be expected to provide the same amount of income to the borough or the state because many of the facilities will be offshore, beyond the reach of local and state taxation. Furthermore, this is a critical component to analyzing, quantifying, and managing potential effects from the ultimate downturn of oil revenues. It is particularly deserving of attention because MMS has influence over whether facilities are located onshore or offshore, and because the effect could vary greatly depending on whether potential revenues to the North Slope Borough, and the implied effects, are considered in the decision of where to locate facilities. In addition, attention needs to be devoted to the degree to which continued oil development, if feasible, would in fact contribute to the long-term well-being of the region and its people or exacerbate the difficulties created by the inevitable longer-term "bust. " The state of Alaska has been able to establish at least a sizable permanent fund from the relatively short-term bonanza extracted from Prudhoe Bay, but the situation for the North Slope Borough might be more challenging; although the borough has set up its own permanent fund, it currently stands at only $259 million. Even with substantial additions during the remaining years of operation at Prudhoe Bay, the fund is likely to prove far too small to support anything like the current level of borough activities on a continuing basis. The situation is further complicated by the fact that the borough's long-term bonded indebtedness as of July 1993 was $728 million.

CONCLUSIONS AND RECOMMENDATIONS

The specific conclusions and recommendations for this chapter follow. For the general and overall conclusions of this report, see Chapter 8.

Conclusion 1: The development of a good data base on gouging rates and the infilling process requires yearly replicate measurements of the gouging patterns on selected representative regions of the Alaskan shelf. Although some such data exist (Barnes et al., 1978; Rearic et al., 1981), the expansion of this data base would be highly desirable and would contribute significantly to safe pipeline design.

Recommendation 1:   *Collect yearly replicate measurements of gouging depths and patterns once an oil field has been identified.* MMS should support a small but long-term (5 years minimum) program that collects annual replicate measurements of gouging depths and patterns from selected, representative regions of the Beaufort and Chukchi shelves. Of particular initial importance now is the offshore area in the vicinity of the Kuvlum discovery in the Beaufort Sea. Although a Kuvlum study will not replace site-specific information from other areas, it should contribute significantly to pipeline route selection and to safe pipeline design. The personnel requirement for this study would be 3 persons per year for each 5-year study.

Alternative: The alternative to adequate temporal data is significant overdesign of the pipeline either by deeper burial than actually required or by armoring.

Conclusion 2: Information is needed on the rate of gouge infilling and on potential scouring problems along pipeline routes as they are identified.

Recommendation 2: *Expand observational data base on the temporal variations in near-sea currents.* MMS should support a program that expands the observational data base on the temporal variations in near-seafloor currents and in sediment transport for both the Beaufort and Chukchi shelves. Measurements should be taken at several water depths at representative locations over a 3-year period. The personnel requirement for this study would be 1½ person-years per year for a project total of 4½ person-years. The deployment and recovery of buoys could be carried out as a part of the study discussed as in Recommendation 1.

Alternative: Estimates might be provided by attempting to model both the currents and the resulting sediment fluxes. However, in the absence of adequate field data, there would be no way to evaluate the adequacy of the estimated values. In that the oceanographic conditions along both the Beaufort and Chukchi coasts are believed to be complex, field data are probably essential. The ultimate consequence of inadequate offshore pipeline design could, of course, be a spill below the pack ice. The social, economic, political, and ecological consequences could be devastating.

Conclusion 3: There is a need for further studies that will contribute to the development of effective remote sensing techniques that are capable of locating oil hidden either within or, more commonly, below sea ice.

Recommendation 3: *Develop remote-sensing techniques to locate oil within or under ice.* Preliminary assessment of currently available technology would require 2 years and a total of 4 person-years of effort, assuming that some of the instrument systems are operated by contractors. The cost of developing a selection of effective systems for detecting oil under ice might be in the range of $3-5 million, including design, construction, and testing. This is a very rough estimate, because no truly promising technologies have yet been identified. Acoustic methods derived from military submarine developments might be promising, now that some are potentially available for development and application in the private sector under cooperative research and development agreements.

Alternative: There are no effective alternatives. Drilling pilot holes in the ice with hand equipment, which could be the only current alternative, is ineffective over a large spill area.

Conclusion 4: It is necessary to integrate all relevant available data sets so that decision makers are fully informed.

Recommendation 4: *Use data sets from the Canadian North, Barents Sea, and Prince William Sound.* Additional integration of data from other places, industry data, data from the Prince William Sound spill, and data from government-sponsored studies (e.g., the Anchorage Symposium on the Prince William Sound Spill sponsored by the National Oceanic and Atmospheric Administration) should be synthesized and evaluated. Also, data released by the Exxon Corporation for the Prince William Sound spill at the American Society for Testing and Materials Atlanta Conference in April 1993 should be evaluated. This effort would require 1 person-year. The neglect of applicable studies that have been carried out at other locations by other concerns is, at best, a waste of both time and money.

Alternative: None recommended.

Conclusion 5: Regardless of probability, oil spills are a major concern in the Arctic because of the perceived or real inability to clean them up or control them. A few of the many techniques used in temperate regions have been tried in experimental tanks with cold water. There is only limited experience in cold and ice-infested waters with experimental spills.

Recommendation 5: *MMS should attempt to evaluate critically the relative effectiveness of different countermeasures as applied under a representative range of conditions that might be expected in the Arctic.* Countermeasures could include in-situ burning, use of cold water dispersants, bioremediation, mechanical recovery, and protective procedures. This study should assist in formulating critical questions that should be answered by appropriate field tests. This study should require 2 person-years of effort, and should be cognizant of work carried out on this theme in Canadian and Norwegian waters.

Alternative: None recommended.

Conclusion 6: Because of the difficulty of getting permits to apply oil to shorelines or to spill oil in open water, most spill effects studies have been part of larger programs to study the effects of accidental spills, or "spills of opportunity." Although much valuable information has been gained, there frequently is no opportunity for rigorous science with adequate control or reference sites to detect the effects of the spill or the control/cleanup methods used. Also, resources needed to conduct proper scientific studies (boats, aircraft) are in demand to perform the cleanup and control. Experimental spills are planned in advance, and can be controlled. Reference sites and the number of replicate samples and plots are determined in advance.

Recommendation 6: *MMS should be a major participant in the design and completion of one or more experimental oil spills in the general area of the Chukchi and Beaufort seas.* We believe that experimental spills are essential because they can contribute to the scientific understanding of processes, and fill data gaps, about the interaction of ice and oil. These tests also can contribute to accurate assessments of the abilities and limits of countermeasures, and remediation and restoration procedures. The data from recent tests should be evaluated. Carefully controlled field tests would be an invaluable extension of such work into a realistic environmental setting. Experimental spills also are essential in the training of local industry and community spill teams. The performance of spill response teams cannot be expected to be excellent if they are never permitted to practice using real oil. Also, preimpact data could not be collected without an experimental spill and would prevent proper evaluation of spill, control, and cleanup effects.

An estimate of the cost of an investigation of an experimental shoreline spill would be $1 million to $1.5 million per year, carried out at years 0, 1, 2, 3, and 5. An open-water or ice spill investigation carried out for 1 year might cost $1 million to $2 million; a 2-year investigation would probably be needed. This activity would require 3 person-years per year for the shoreline investigation, and 4 person-years per year for an open water-ice spill trial.

Alternative: None recommended.

Conclusion 7: Outside the U.S., the individual agencies and companies involved in arctic exploration and development are conducting experimental spill studies with little interaction with MMS.

Recommendation 7: *Institute cooperative research planning and funding of experimental spill studies.* We generally recommend fewer, more comprehensive projects rather than a large number of small projects. Spills should be used for a variety of purposes simultaneously (research, evaluation of countermeasures technology, training) and supported cooperatively by federal and state agencies and industry.

Alternative: None recommended.

Conclusion 8: It is essential that EISs be current with state and federal legislation.

Recommendation 8: *Ensure that EISs reflect the new federal and state legislation.* Regulatory requirements greatly influence requirements for oil spill preparedness and response. This would require ½ person-year.

Conclusion 9: MMS's approach to mitigation and remediation project selection has largely been from the "bottom up." Useful information has been obtained from these studies. MMS should now focus its mitigation and remediation activities on the most critical potential sources of environmental damage. The discussion in Chapter 6 clearly identifies that traditional knowledge should be used in study design and consulted for lease decisions.

Recommendation 9: *Take a top-down approach, i.e., an approach in which an overall scientific vision or framework influences the* *studies program, in selecting mitigation and remediation studies to be conducted.* It is now opportune for MMS to make a transition to "top-down" studies that are critically targeted to specific industry activities. This approach should optimize the chances that, with limited funding, support will be given to projects that would contribute significantly to environmentally safe, efficient, and timely development of offshore resources. These studies should be integrated with continuing long-term studies of Arctic ecosystems and processes. There should be no incremental resource requirement to implement this recommendation. It will be more cost effective to fund lower-level, long-term studies using disciplinary experts than to respond sequentially to crises. This latter approach is financially exorbitant, politically costly, and does not create a positive working relationship between agencies and the affected public.

Alternative: Research will be in response to crises and the resultant high-cost, short-term studies will be relatively ineffective.

Conclusion 10: Most of MMS's research on oil spills associated with lease sales has focused on documenting potential effects rather than on preventing them or minimizing their consequences. MMS does have a program on countermeasures that is not associated with its lease sale programs, and this program has contributed valuable information about in-situ burning and the use of dispersants and other chemical agents.

Recommendation 10: *Develop specific countermeasures to minimize the effects of arctic spills.* The next step is to look at particular lease-sale areas and develop specific countermeasures. More research is needed on in-situ burning on ice, in ice, and on cold water. The ecological tradeoffs involved in using cold-water dispersants need to be evaluated and compared with other spill control technologies in the Arctic. The effectiveness of mechanical recovery systems should be evaluated under arctic conditions. Shoreline cleanup and remediation, including bioremediation methods, also should be studied. This effort, shared with other organizations, might have a cost to MMS of $5 million and require 3-5 years to implement.

Alternative: None recommended.

Conclusion 11: Gaps still exist in fate, behavior, and effects data bases; therefore, the accuracy of models used is dubious.

Recommendation 11: *Fate, behavior, and effects studies are needed both to fill data gaps for arctic environments and to improve or verify models.* Experimental spills are the preferred way to acquire this information, but spills of opportunity also should be studied as appropriate and a science response plan should be developed to take advantage of the study opportunities they provide. More work is needed on developing and verifying models that assess fate, behavior, and effects of spills. Models should always be compared with field data and refined accordingly. The financial cost of this effort might be $1 to $1.5 million; duration cannot be estimated because the committee cannot know when the opportunity might arise.

Alternative: None recommended.

Conclusion 12: MMS has not fully apprised North Slope Borough communities and oil industry operators of its research, nor has it encouraged their participation in development of research programs.

Recommendation 12: *North Slope Borough communities and oil industry operators should be kept informed about the progress of MMS research on oil spill response and other scientific studies and, as much as possible, participate in the development of research programs.*

Alternative: None recommended.

Conclusion 13: The Alaska OCS oil and gas leasing process suffers from the same sort of environmental gridlock found in other regions.

Recommendation 13: *MMS needs to review studies on "environmental gridlock" and incorporate lessons from these studies into the decision-making process.* These studies demonstrate that development of a process based on mutual trust is an integral component of the solution. This process requires a genuine sharing of power, a sincere effort to seek accommodation with local groups, and should begin with a recognition that many answers about lease-stage and development-stage impacts are not yet known.

If properly implemented, such an approach could permit MMS to work cooperatively with affected groups in a good-faith effort to develop relevant answers where possible and to develop mutually acceptable procedures for dealing with residual and unavoidable uncertainties where answers are not likely within a reasonable period and at reasonable cost.

Alternative: None recommended.

8 GENERAL CONCLUSIONS AND AN ALTERNATIVE TO ADDITIONAL STUDIES

The general conclusions about the adequacy of scientific and technical information relevant to the potential environmental consequences of the lease-sale areas follow. (Refer to each chapter for discipline-specific conclusions and recommendations.) This chapter also discusses an alternative to conducting additional studies in a case in which it appears that additional information alone will not permit a resolution of disagreements.

GENERAL CONCLUSIONS

Conclusion 1: The committee concludes that the environmental information available for the Chukchi, Navarin, and Beaufort OCS areas is generally adequate for leasing and exploration decisions, except in the case of effects on the human environment (i.e., socioeconomic effects, as defined in the Outer Continental Shelf Lands Act (OCSLA)). Prelease and lease-stage effects on the human environment—which can begin even before any physical or biological changes take place—are a special category of effects and are discussed in Chapter 6. In general, the information available for resource geology, the physical environment, biotic resources, spills, and mitigation and remediation activities adequately reflects the differences between arctic OCS areas and other U.S. OCS areas where development and production have already occurred.

Conclusion 2: There is a considerable range in the amount of data needed to provide adequate information for decisions regarding development, production, transportation, siting of onshore and offshore facilities, and termination of activities. In general, site-specific studies of biotic resources, especially marine mammals critical to the subsistence economy of Alaska Natives, of physical conditions, and of long-term effects on Native and non-Native residents, will be necessary to determine acceptable risks to human and other living resources prior to development and production. The proper sites for these studies can be determined only by exploration. They should be carried out to support the various permitting processes following leasing decisions.

Conclusion 3: Information from research programs in other nations with conditions similar to the Alaskan arctic OCS and from the experience base of Alaska Natives has not been fully exploited. More coordination of government (local, state, federal), industry, and Native information and collaboration among these groups are needed to promote efficiency in study efforts and an improved atmosphere of mutual trust.

Conclusion 4: MMS's Environmental Studies Program and oil and gas resource assessment efforts have yielded credible information useful for establishing a general baseline or characterization of the living resources, physical conditions, social and economic setting, and potential oil and gas resources in the arctic OCS. The committee concludes that MMS should now concentrate its resources on fewer, longer-term studies of the living resources, social and economic conditions, and physical processes with peer review of results, to develop the additional information needed for decision making about oil and gas production, transportation, and related development.

Conclusion 5: The most controversial of the potential effects of OCS development in the arctic OCS are oil spills and interference with marine mammals—especially the bowhead whale—critical to the subsistence economy of Alaska Natives. These issues must be dealt with and Alaska Natives and their experts should be substantially involved, otherwise plans to develop arctic OCS oil and gas resources could be seriously jeopardized.

Conclusion 6: The committee suggests that it could be time for the Office of Management and Budget's (OMB's) criteria to be re-evaluated. OCSLA explicitly requires that "Subsequent to the leasing and developing of any region, the Secretary *shall* conduct such additional studies as he deems necessary and *shall* monitor the human, marine, and coastal environments . . . in a manner designed to provide time-series and data trend information which can be used for comparison with any previously collected data for the purpose of identifying any significant changes in the quality and productivity of such environments" (43 U.S.C. § 1346(c)) (emphasis added).

Although the committee is aware that MMS has its funding approved by OMB—by the "A2B1" criteria—the committee is concerned that the criteria could present a significant obstacle that might not have been fully appreciated when they were first approved. The need for long-term post-leasing and post-development studies as required by OCSLA cannot be met unless enough money is available to do the studies.

AN ALTERNATIVE TO ADDITIONAL STUDIES

The traditional approach to facility siting is to use scientific methods to evaluate possible sites with a view toward potential impacts—most often through the EIS process—and to employ technical knowledge for avoiding or reducing them. This is more or less the paradigm established by OCSLA and embodied in the Environmental Studies Program and the OCS oil and gas leasing process. However, the committee notes that OCSLA specifically includes a "congressional declaration of policy" that states and affected local governments "are entitled to an opportunity to participate, to the extent consistent with the national interest, in the policy and planning decisions made by the Federal Government relating to exploration for, and development and production of, minerals of the Outer Continental Shelf" (43 U.S.C. § 1332 (4)(B)). The law further calls for the secretary to carry out environmental studies "in full cooperation with" affected states (43 U.S.C. § 1346(c)).

It is becoming increasingly difficult to argue that a centralized approach to siting undesirable facilities is in the national interest. The approach is frequently met with almost overwhelming local opposition. Moreover, this approach creates or exacerbates effects, as is reflected in the discussion of environmental gridlock in Chapter 6. Although additional studies could satisfy the OCSLA mandate for studying impacts on the human environment, they will not likely satisfy communities that leasing is in their best interest, nor will additional studies, in themselves, allow MMS to fulfill the OCSLA mandate to develop the information that will be used to manage impacts to the human environment. MMS has not yet developed a decision-making process that is broadly acceptable to the communities of northern Alaska. Social scientists have extensively studied processes that are conducive to cooperative solutions (O'Hare et al., 1983; Gregory et al., 1991), and several of their conclusions appear to be directly applicable. No process can guarantee success, but it is becoming increasingly clear that the federal government's current methods almost guarantee failure to arrive at a cooperative solution.

An alternative approach would require a significant change from current ways of doing business. The most important factor for achieving a cooperative solution to facility siting is an established basis for trust among the parties, founded largely on three considerations. The first is the most common but least formal, and it is perhaps the least relevant in the present case: Many forms of trust require a track record of experience, during which the parties involved have observed one another's actions over a long enough time to have concluded that everyone involved has behaved in a manner consistently that is deserving of trust. Where no such record is available, or worse, when the participants believe they have reason to be actively suspicious of one another, the second and third considerations come into play.

The second consideration is that decisions must be based on a process that is viewed as sound, fair, and acceptable to all parties. As stated by Gregory et al. (1991), "If the assessment process is viewed as flawed or biased by political considerations, then no siting agreement will be reached." The third consideration is that the community must have real control over decisions that influence risks. This means that communities must have an active part in the process. If the community believes that its concerns are ignored or discounted, it will almost certainly distrust the process.

Unfortunately, a simplistic approach—such as the apparently straightforward exhortation that the various parties involved should seek to use cooperation as a means of dealing with uncertainties— tends to work only in cases having the first of the three considerations noted above. This approach would appear, on the basis of past studies, to have little likelihood of success in the present case. Under present circumstances, many of the parties involved tend to view one another with distrust. It is possible for trust to grow over time, but the process is almost always slow, and to work things out cooperatively, trust must be present in advance. As noted by Slovic et al. (1991), there tends to be an "asymmetry principle" at work: It is far easier to lose trust than it is to gain it.

Historical Context for Mistrust

Although we do not believe that the current position is completely hopeless, we are not optimistic for several reasons that the historical context provides a solid basis for trust among the parties. First, there appear to be strong feelings among those on the Alaska North Slope that the process is flawed and is driven by political considerations. Indeed, this committee has learned that the MMS program in Alaska shares with many other programs the problem of a serious credibility deficit, whether or not its environmental studies are credible in an objective sense.

Second, the experience with the *Exxon Valdez* oil spill also contributes to a lack of credibility in the potential to clean up spilled oil effectively, even though that accident did not involve OCS oil and did not physically affect northern Alaska. Major changes have taken place since the accident, such as stockpiles of cleanup and control equipment, complete revision of the contingency plans by federal, state, and local government and Alyeska, and many drills and training exercises are conducted with each company, state and federal agencies, and locals. It is hoped that these changes will enhance future response effectiveness. To some extent, it is unreasonable to expect any effort of the magnitude of that cleanup, put into the field so quickly, to have been more precisely choreographed, despite the slow response and confusion in the earliest stages of the spill. It is just not possible to prevent all impacts from very large spills.

The *Exxon Valdez* spill does, however, reflect the reality of experience in Alaska, where many residents also remember earlier repeated assertions about safety, preparedness, and even double-hulled tankers, along with other claims that were made in the heat of political debates over the Trans-Alaska Pipeline System. These factors lead to an erosion of credibility and trust among the parties, which aggravates impacts and diminishes the potential for cooperative solutions. The limitations on our ability to respond immediately and completely successfully to a major accident should be acknowledged.

Negotiation Process

Another factor in the historical context is a different kind of asymmetry principle, which deals with differential levels of influence in the negotiating process. In most of the several cases where parties have successfully negotiated—even in cases in which the parties involved have been openly antagonistic or hostile toward one another—the parties were able to exercise at least a rough parity in negotiating leverage. This is possible because neither party is required to trust another. In some cases, a negotiated agreement can be overseen by a third party that is trusted by both sides, or it might be possible for one side to impose penalties if an agreement fails to be carried out. However, in cases where the distrusted party is a government agency, few alternatives exist that do not require a genuine and tangible sharing of decision-making power with the affected parties.

Several affected parties told the committee that they consider circumstances to have become relatively polarized and that they do not see MMS as a neutral third party—independent both of the oil industry and of local and environmental concerns. Instead, they see MMS as allied with industrial and developmental interests. These local perceptions can be debated, but they have real consequences, which include the fact that MMS is not likely to be trusted as an honest broker.

Under the circumstances, the affected parties widely believe their best or only real opportunity to influence the course of development comes before the leasing decision. Once a lease is issued, if commercially valuable quantities of gas or oil are discovered, residents are not likely to have significant influence over the course of development. Some MMS personnel with whom the committee discussed this issue disagreed strongly with this assessment; others largely agreed. None of them was able to offer specific examples of cases in which local objections had led the agency to make truly significant changes in development plans where the leases had been offered and commercially significant quantities of oil or gas had been discovered.

Workable Negotiation Options

Where there is little reason to expect success from exhortations that one party should simply trust the other, there appear to be only two remaining options. One is to continue to perform studies until all of the relevant concerns have been fully explored. In the Arctic, unfortunately, this would be prohibitively expensive, even in light of the large quantities of oil that are now believed to lie offshore. The second option is to develop mechanisms that do not require one party to answer questions in advance to the complete satisfaction of another—that is, through cooperative or negotiated statements. This option also does not require the parties to trust one another in advance.

For cooperative approaches to work, two additional changes are likely to be needed, in addition to increased attention to the scientific credibility of studies as outlined above. First, there will need to be a change in the incentive structure currently in place. Under the current system, the parties who have concerns about development have little faith in their ability to alter the process involved in oil development and production if oil is found. Although the OCS Lands Act provides for cancellation of a lease based on environmental reasons that arise during performance of the lease and compensation for such losses plus interest to the lessee (43 U.S.C. § 1334 (a)(2)(c)), the provision has not been used to date. Thus, those who have concerns about development often conclude that they should oppose even the initial offering of leases.

In such a case, an alternative approach is needed that provides the affected parties with greater control after leases are issued. This would be similar to a number of waste-management controversies in which arrangements have been made for the community to shut down a site if it determines that the facility is being operated unsafely. The most efficient course of action available to MMS at this time could be to modify the existing stipulation mechanism that allows formal input from other agencies so that it also includes more effective input from the affected local communities. Therefore, the affected parties will no longer have to accept on faith that MMS will protect local interests in the long run.

Specific stipulations will need to be worked out by the affected parties. Examples could include requirements that state and local officials approve continuing activities before they can proceed at several points in the development process—when exploration plans are filed, when development plans are filed, after the first year of production, after the fifth and twelfth years of production, and so on. The stipulations, which should allow some degree of predictability for industry, should be part of the initial offering of leases, so that the companies are fully informed in advance, allowing them to account in their bids for the potential risk of being required to deal with additional local concerns. This would give all parties an additional incentive to build and maintain cooperative relationships with local leaders. The additional risk of such stipulations to industry should be low if industry can demonstrate that development can be carried out in accordance with the highest standards of protection of the human, marine, and coastal environments and that the resultant activities can result in significant, long-term economic benefits for affected populations.

Second, even if such stipulations are put into place, a climate of mutual respect will be needed. This will require, in particular, that every effort be made to avoid creating or exacerbating the suspicion that the agency is attempting to dismiss, rather than to deal with, the concerns that are raised by local critics.

A Suggested Example—Bowhead Whales

One of the most controversial aspects of Alaskan OCS oil and gas development concerns the effect of those activities on the migration patterns of marine mammals, especially bowhead whales. Because hunting bowheads is so important to North Slope communities, there have been many studies of the effects of noise on bowhead behavior and migration. Despite those studies, the question is not resolved and it is not clear whether any amount of research will resolve it. For this reason, the committee believes that the best (and perhaps the only) solution is for MMS, the industry, and North Slope residents to attempt to reach agreement on the controversial matters and how they should be adjusted, remedied, or mitigated—as specific times and places that various activities occur—in lieu of or concurrent with additional studies. Such a negotiation does not require any party to give up any rights to other remedies as a condition to begin the negotiation, although the assumption must be that a mutual agreement would be exclusive; the agreed-on course of action will be adhered to unless additional information shows to everyone's satisfaction that it should be modified. If additional studies are conducted, they should be cooperatively designed and implemented by all three parties.

There is no guarantee that this approach would succeed. Nonetheless, it seems unlikely that it could be less successful or more costly than the current system of dueling studies and reviews and their accompanying delays and ill feelings. The approach should work not only for bowhead whales, but for many other areas of controversy questions in Alaska and elsewhere (e.g., the effectiveness of causeway breaches in allowing fish migration). Several practitioners have developed considerable skill in this type of dispute resolution.

Finally, it is of the utmost importance to build effective and thoughtful monitoring programs into any such settlement. Monitoring is essential to evaluate the success of agreements and to provide a basis for facilitating similar agreements elsewhere.

Mitigating Long-Term Socioeconomic Effects

Among the long-term non-spill socioeconomic impacts that need to be dealt with are the potential for cultural erosion and for socioeconomic overadaptation. The potential for overadaptation is exacerbated by the region's remoteness and the limited likelihood of successful economic diversification. Among the obvious possibilities for mitigating those foreseeable effects (as well as for helping to create more positive effects) would be the creation of trust funds. Working cooperatively with the state and affected local governments (including Northwest Alaska Native Association and the North Slope Borough), MMS should explore the potential for mitigating longer-term effects through revenue sharing, as well as through other steps that could help mitigate the coming "bust" by building up locally controlled trust funds. If such funds were sufficiently large, they could even help cushion the impending end of the Prudhoe Bay revenues, with the principal being left intact, and with the annual proceeds being used to fund a significant fraction of the borough's (or NANA's) current employment and other costs. Revenue sharing should be used for mitigating regional socioeconomic effects. It should not be used for mitigating effects that are of national concern.

References

Aagaard, K. 1984. The Beaufort undercurrent. Pp. 47-71 in The Alaskan Beaufort Sea: Ecosystems and Environment, P.W. Barnes, D.M. Schell and E. Reimnitz, eds. New York: Academic Press.

Aagaard, K., and P. Griesman. 1975. Toward new mass and heat budgets for the Arctic Ocean. J. Geophys. Res. 80:3821-3827.

Aagaard, K., L.K. Coachman, and E.C. Carmack. 1981. On the halocline of the Arctic Ocean. Deep-Sea Res. 28:529-545.

Aagaard, K., C.H. Pease, A.T. Roach, and S.A. Salo. 1989. Beaufort Sea Mesoscale Circulation Study: Final Report. NOAA Tech. Memo. ERL PMEL-90, Pacific Marine Environmental Laboratory, Seattle, Wash. 114 pp.

Afanas'ev, V.P., Iu.V. Dolgopolov, and Z.I. Shraishtein. 1973. Ice pressure on individual marine structures. Pp. 50-68 in Studies in Ice Physics and Ice Engineering, G.N. Yakovlev and R. Hardin, eds. National Science Foundation, Washington, D.C.

Alaska Clean Seas. 1983. Alaskan Beaufort Sea Coastal Region, Vol. 2. Biological Resources Atlas. Alaska Clean Seas, Anchorage, Alaska.

Alaska Department of Fish and Game. 1977. Biophysical Boundaries for Alaska's Coastal Zone. Marine Coastal Habitat Management Program, Fairbanks, Alaska. 59 pp.

Allen, A.A. 1988. Comparison of response options for offshore oil spills. Pp. 289-306 in Proceedings of the Eleventh Arctic and Marine Oil Spill Program (AMOP) Technical Seminar, Vancouver, British Columbia, June 12-14. Environment Canada, Ottawa, Ontario.

Alverson, D.L., and N.J. Wilimovsky. 1966. Fishery investigations of the southeastern Chukchi Sea. Pp. 843-860 in Environment of the Cape Thompson Region, Alaska, N.J. Wilimovsky and J.N. Wolfe, eds. U.S. Atomic Energy Commission, Oak Ridge, Tenn.

American Fisheries Society, Committee on Names of Fishes. 1991. Common and Scientific Names of Fishes from the United States and Canada, Special Publ. 20, 5th Ed., C.R. Robins, chair. American Fisheries Society, Bethesda, Md.

Amstrup, S.C., I. Stirling, and J.W. Lentfer. 1986. Past and present status of polar bears in Alaska. Wildl. Soc. Bull. 14:241-254.

Andrew, J.M., and J.C. Haney. 1993. Water masses and seabird distributions in the southern Chukchi Sea. Pp. 381-388 in Results of the Third Joint US-USSR Bering and Chukchi Seas Expedition (BERPAC), Summer 1988, P.A. Nagel, ed. U.S. Fish and Wildlife Service, U.S. Department of the Interior, Washington, D.C.

API (American Petroleum Institute) Task Force. 1985. Oil Spill Response: Options for Minimizing Adverse Ecological Impacts. Publ. No. 4398. American Petroleum Institute, Washington, D.C. 98 pp.

ARC (U.S. Arctic Research Commission). 1991. Goals, Objectives and Priorities to Guide United States Arctic Research. No. 7 (Jan.):23. U.S. Arctic Research Commission, Washington, D.C.

ARCO Alaska, Inc. 1989. ARCO Arctic Environmental Reports Collection, Catalog and Index. Arctic Environmental Information and Data Center (AEIDC), University of Alaska, Anchorage.

Assur, A. 1961. Traffic over frozen or crusted surfaces. Pp. 913-923 in Proceedings of the First International Conference on the Mechanics of Soil-Vehicle Systems. Torino-St. Vincent, Italy.

ASTM (American Society for Testing and Materials). 1986. Guide for Ecological Considerations for the Use of Chemical Dispersants in Oil Spill Response: The Arctic. Standard F 1012-86. American Society for Testing and Materials Standard, Philadelphia, Pa.

ASTM (American Society for Testing and Materials). 1993. Third Symposium on Environmental Toxicology and Risk Assessment: Aquatic, Plant, and Terrestrial. ASTM Committee E-47 on Biological Effects and Environmental Fate, April 26-29, Atlanta, Ga. American Society for Testing and Materials, Philadelphia, Pa.

Bailey, A.M. 1948. Birds of Arctic Alaska. Popular Series 8. Colorado Museum Natural History, Denver, Colo.

Baker, J.M., R.B. Clark, P.F. Kingston, and R.H. Jenkins. 1990. Natural

Recovery of Cold Water Marine Environments after an Oil Spill. 13th Arctic and Marine Oilspill Program Technical Seminar, June. Environmental Emergencies Technologies Division, Ottawa, Ontario. 111 pp.

Baldassare, M. 1985. Trust in local government. Social Sci. Q. 66:704-712.

Barnes, P.W. 1979. Fast-Ice Thickness and Snow Depth in Relation to Oil Entrapment Potential, Prudhoe Bay, Alaska, USGS Open File Rep. 79-539. U.S. Geological Survey, Denver, Colo. 30 pp.

Barnes, P.W., and E. Reimnitz. 1974. Sedimentary processes on arctic shelves off the northern coast of Alaska. Pp. 439-476 in The Coast and Shelf of the Beaufort Sea, J.C. Reed and J.E. Sater, eds. Arctic Institute of North America, Arlington, Va.

Barnes, P.W., and E. Reimnitz. 1979. Ice Gouge Obliteration and Sediment Redistribution Event, 1977-1978, Beaufort Sea, Alaska. USGS Open File Rep. 79-848. U.S. Geological Survey, Denver, Colo. 22 pp.

Barnes, P.W., D.M. McDowell, and E. Reimnitz. 1978. Ice Gouging Characteristics: Their Changing Patterns from 1975-1977, Beaufort Sea, Alaska. USGS Open File Rep. 78-730. U.S. Geological Survey, Denver, Colo. 42 pp.

Barnes, P.W., D.M. Rearic, and E. Reimnitz. 1984. Ice gouging characteristics and processes. Pp. 185-212 in The Alaskan Beaufort Sea: Ecosystems and Environment, P.W. Barnes, D.M. Schell, and E. Reimnitz, eds. Toronto, Ontario: Academic Press.

Barthelemy, J.L. 1990. Thirty-five years of sea ice runway operation—McMurdo Station, Antarctica. Pp. 864-877 in IAHR 10th Symposium on Ice, Vol. 2. Helsinki University of Technology, Espoo, Finland.

Blenkarn, K.A. 1970. Measurement and Analysis of Ice Forces on Cook Inlet Structures. Paper No. OTC-1261. 1970 Proceedings of the Annual Offshore Technology Conference. Offshore Technology Conference, Houston, Tex.

Blomquist, S., and M. Elander. 1981. Sabine's gull (*Xema sabini*), Ross' gull (*Rhodostethia rosea*), and Ivory gull (*Pagophila eburnea*): Gulls in the Arctic: A review. Arctic 34:122-132.

Bloom, G.L. 1964. Water transport and temperature measurements in the eastern Bering Strait, 1953-1958. J. Geophys. Res. 69:3335-3354.

Boaz, I.B., and D.N. Bhula. 1981. A steel production structure for the

Alaskan Beaufort Sea. In 1981 Proceedings of the Annual Offshore Technology Conference. Offshore Technology Conference, Houston, Tex.

Boehm, P.D. 1987. Transport and transformation processes regarding hydrocarbon and metal pollutants in offshore sedimentary environments. Pp. 233-286 in Long-Term Environmental Effects of Offshore Oil and Gas Development, D.F. Boesch and N.N. Rabalais, eds. New York: Elsevier Applied Science.

Boesch, D.F., and N.N. Rabalais, eds. 1987. Long-Term Environmental Effects of Offshore Oil and Gas Development. New York: Elsevier Applied Science.

Bowles, R.T. 1981. Preserving the contribution of traditional local economies. Hum. Serv. Rural Environ. 6(1):16-21.

BP Exploration (Alaska), Inc. 1991. Industry-Sponsored Research on Alaska's North Slope. Environmental and Regulatory Affairs, BP Exploration (Alaska), Inc., Anchorage, Alaska.

Bradstreet, M.S.W. 1980. Thick-billed murres and black guillemots in the Barrow Strait area, N.W.T., during spring: Diets and food availability along ice edges. Can. J. Zool. 58:2120-2140.

Bradstreet, M.S.W., and W.E. Cross. 1982. Trophic relationships at high arctic ice edges. Arctic 35:1-12.

Bradstreet, M.S.W., K.J. Finley, A.D. Sekerak, W.B. Griffiths, C.R. Evans, M.F. Fabijan, and H.E. Stallard. 1986. Aspects of the Biology of Arctic Cod (*Boreogadus saida*) and its Importance in Arctic Marine Food Chains. Canadian Technical Report of Fisheries and Aquatic Sciences No. 1491. Central and Arctic Region, Department of Fisheries and Oceans, Winnipeg, Manitoba. 193 pp.

Braham, H.W., J.J. Burns, G.A. Fedoseev, and B.D. Krogman. 1984. Habitat partitioning by ice-associated pinnipeds: Distribution and density of seals and walruses in the Bering Sea, April 1976. Pp. 25-48 in Soviet-American Cooperative Research on Marine Mammals: Pinnipeds, Vol. 1, F.H. Fay and G.A. Fedoseev, eds. NOAA Tech. Rep. NMFS 12. National Oceanic and Atmospheric Administration, U.S. Department of Commerce, Washington, D.C. 104 pp.

Brower, W.A., H. Diaz, A. Prechtel, H. Searby, and J. Wise. 1977a. Climatic Atlas of the Outer Continental Shelf Waters and Coastal Regions of Alaska: Chukchi-Beaufort Sea, Vol. 3. National Oceanic and Atmospheric Administration/Outer Continental Shelf Environmental

Assessment Program (NOAA/OCSEAP) Final Report. Arctic Environmental Information and Data Center, Anchorage, Alaska.

Brower, W.A., H. Diaz, A. Prechtel, H. Searsby, and J. Wise. 1977b. Climatic Atlas of the Outer Continental Shelf Waters and Coastal Regions of Alaska: Bering Sea, Vol. 2. National Oceanic and Atmospheric Administration/Outer Continental Shelf Environmental Assessment Program (NOAA/OCSEAP) Final Report. Arctic Environmental Information and Data Center, Anchorage, Alaska.

Brown, W.Y. 1984. Arctic environmental quality. Pp. 178-198 in United States Arctic Interests: The 1980s and 1990s, W.E. Westermeyer and K.M. Shusterich, eds. New York: Springer-Verlag.

Brown, H.M., and R.H. Goodman. 1986. In-situ burning of oil in ice leads. Pp. 245-256 in Proceedings of the 9th Arctic and Marine Oilspill Program Technical Seminar. Environment Canada, Ottawa, Ontario.

Brueggeman, J.J., and R.A. Grotefendt. 1984. Seal, sea lion, and walrus surveys of the Navarin Basin, Alaska. National Oceanic and Atmospheric Administration/Outer Continental Shelf Environmental Assessment Program (NOAA/OCSEAP) Final Reports of Principal Investigators 37(1986):1-94.

Brueggeman, J.J., R.A. Grotefendt, and A.W. Erickson. 1984. Endangered whale surveys of the Navarin Basin, Alaska. National Oceanic and Atmopsheric Administration/Outer Continental Shelf Environmental Assessment Program (NOAA/OCSEAP) Final Reports of Principal Investigators 42(1986):1-146.

Bunker, S.G. 1984. Modes of extraction, unequal exchange, and the progressive underdevelopment of an extreme periphery: The Brazilian Amazon, 1600-1980. Am. J. Sociol. 89:1017-1064.

Burns, J.J. 1970. Remarks on the distribution and natural history of pagophilic pinnipeds in the Bering and Chukchi seas. J. Mammal. 51:445-454.

Burns, J.J. 1981a. Ribbon seal: *Phoca fasciata*. Pp. 89-109 in Handbook of Marine Mammals: Seals, Vol. 2, S.H. Ridgway and R.J. Harrison, eds. New York: Academic Press.

Burns, J.J. 1981b. Bearded seal: *Erignathus barbatus*. Pp. 145-170 in Handbook of Marine Mammals: Seals, Vol. 2, S.H. Ridgway and R.J. Harrison, eds. New York: Academic Press.

Burns, J.J. 1984. Living resources. Pp. 75-104 in United States Arctic Interests: The 1980s and 1990s, W.E. Westermeyer and K.M.

Shusterich, eds. New York: Springer-Verlag.

Burns, J.J., and B.P. Kelly. 1982. Studies of Ringed Seals in the Alaskan Beaufort Sea During Winter: Impacts of Seismic Exploration. Annual report to National Oceanic and Atmospheric Administration/Outer Continental Shelf Environmental Assessment Program (NOAA/OCSEAP), Anchorage, Alaska. 57 pp.

Burns, J.J., and G.A. Seaman. 1986. Investigations of Beluga whales in coastal waters of western and northern Alaska: II. Biology and ecology. National Oceanic and Atmospheric Administration/Outer Continental Shelf Environmental Assessment Program (NOAA/OCSEAP) Final Reports of Principal Investigators 56(1988):221-357.

Burns, J.J., L.H. Shapiro, and F.H. Fay. 1981. The relationships of marine mammal distributions, densities and activities to sea ice conditions. Environmental Assessment of the Alaskan Continental Shelf, Final Reports, Biological Studies 11:489-670.

Cammaert, A.B., and D.A. Muggeridge. 1988. Ice Interaction with Offshore Structures. New York: Van Nostrand Reinhold.

Carmack, E.C. 1990. Large-scale physical oceanography of polar oceans. Pp. 171-222 in Polar Oceanography. Part A: Physical Science, W.O. Smith, Jr., ed. New York: Academic Press.

Carmack, E.C., R.W. Macdonald, and J.E. Papadakis. 1989. Water mass structure and boundaries in the Mackenzie shelf/estuary. J. Geophys. Res. 94:18043-18055.

Carman, G.J., and P. Hardwick. 1983. Geological and regional setting off Kuparuk oil field, Alaska. Am. Assoc. Petrol. Geol. Bull. 67:1014-1031.

Centaur Associates. 1986. Indicators of the Direct Econmic Impacts Due to Oil and Gas Development in the Gulf of Mexico—Executive Summary. MMS 86-0016. Final report to the Gulf of Mexico OCS Region, Minerals Management Service, U.S. Department of the Interior, New Orleans, La.

Chance, N. 1990. The Inupiat and Arctic Alaska: An Ethnography of Development. Fort Worth, Tex.: Holt, Rinehart and Winston.

Chen, A.C.T., and J. Lee. 1986. Large-scale ice strength tests at slow strain rates. Pp. 374-378 in Proceedings of the Fifth International Offshore Mechanics and Arctic Engineering (OMAE) Symposium, Vol. 4. American Society of Mechanical Engineers, New York.

Chénard, P.G., F.R. Engelhardt, J. Blane, and D. Hardie. 1989. Patterns of oil-based drilling fluid utilization and disposal of associated wastes on the Canadian offshore frontier lands. Pp. 119-136 in Drilling Wastes, F.R. Engelhardt, J.P. Ray, and A.H. Gillam, eds. New York: Elsevier Applied Science.

Clark, R.C., Jr., and J.S. Finley. 1977. Effects of oil spills in arctic and subarctic environments. Pp. 411-466 in Effects of Petroleum on Arctic and Subarctic Marine Environments and Organisms: Biological Effects, Vol. 2, D.C. Malins, ed. New York: Academic Press.

Clark, R.C., Jr., and J.S. Finley. 1982. Occurrence and impact of petroleum on arctic environments. Pp. 295-341 in The Arctic Ocean: The Hydrographic Environment and the Fate of Pollutants, L. Rey, ed. New York: Wiley-Interscience.

Clarkson, P.L., P.A. Gray, J.E. McComiskey, L.R. Quaife, and J.G. Ward. 1986. Managing polar bear problems in northern development areas. In Northern Hydrocarbon Development Environmental Problem Solving: Proceedings of the Eighth Annual Meeting of the International Society of Petroleum Industry Biologists, Sept. 24-26, 1985, Banff, Alberta, M.L. Lewis, ed. Toronto, Ontario: University of Toronto Press.

Coachman, L.K. 1986. Circulation, water masses and fluxes on the southeastern Bering Sea shelf. Continental Shelf Res. 5:23-108.

Coachman, L.K., and K. Aagaard. 1981. Revaluation of water transports in the vicinity of Bering Strait. Pp. 95-110 in The Eastern Bering Sea Shelf: Oceanography and Resources, Vol. 1, D.W. Hood and J.A. Calder, eds. National Oceanic and Atmospheric Administration, Washington, D.C.

Coachman, L.K., and K. Aagaard. 1988. Transport through the Bering Strait: Annual and interannual variability. J. Geophys. Res. 93:15535-15539.

Coachman, L.K., K. Aagaard, and R.B. Tripp. 1975. Bering Strait: The Regional Physical Oceanography. Seattle: University of Washington Press. 172 pp.

COGLA (Canada Oil and Gas Lands Administration). 1985a. Effects of Explosives in the Marine Environment. Tech. Rep. No. 6. Canada Oil and Gas Lands Administration, Ottawa, Ontario.

COGLA (Canada Oil and Gas Lands Administration). 1985b. Relief Well Drilling Capability on Canada Lands. Tech. Rep. No. 4. Canada Oil and Gas Lands Administration, Ottawa, Ontario.

Colony, R. 1985. A Markov model for nearshore sea ice trajectories. National Oceanic and Atmospheric Administration/Outer Continental Shelf Environmental Assessment Program (NOAA/OCSEAP) Final Reports of Principal Investigators 72(1990):1-55.

Colson, E. 1971. The Social Consequences of Resettlement: The Impact of the Kariba Resettlement upon the Gwembe Tonga. Manchester, U.K.: Manchester University Press.

Connors, P.G. 1984. Ecology of shorebirds in the Alaskan Beaufort littoral zone. Pp. 403-416 in The Alaskan Beaufort Sea: Ecosystems and Environments, R. Barnes, D.M. Shell, and E. Reimnitz, eds. New York: Academic Press.

Connors, P.G., and C.S. Connors. 1982. Shorebird littoral zone ecology of the southern Chukchi coast of Alaska. National Oceanic and Atmospheric Administration/Outer Continental Shelf Environmental Assessment Program (NOAA/OCSEAP) Final Reports of Principal Investigators 35(1985):1-57.

Connors, P.G., J.P. Myers, and F.A. Pitelka. 1979. Seasonal habitat use by Alaskan arctic shorebirds. Stud. Avian Biol. 2:101-111.

Connors, P.G., S.R. Johnson, and G.J. Divoky. 1981a. Birds. Pp. 39-42 in Ecological Characterization of the Sale 71 Environment, Beaufort Sea (Sale 71) Synthesis Report, S.R. Johnson, ed. National Oceanic and Atmospheric Administration, Juneau, Alaska.

Connors, P.G., C.S. Connors, and K.G. Smith. 1981b. Shorebird littoral zone ecology of the Alaskan Beaufort coast. National Oceanic and Atmospheric Administration/Outer Continental Shelf Environmental Assessment Program (NOAA/OCSEAP) Final Reports of Principal Investigators 23(1984):295-396.

Cook, J. 1988. Not in anybody's backyard. Forbes 28:172-182.

Coyle, K.O, G.L. Hunt, Jr., M.B. Decker, and T.J. Weingartner. 1992. Murre foraging, epibenthic sound scattering, and tidal advection over a shoal near St. George Island, Bering Sea. Mar. Ecol. Prog. Ser. 88:1-14.

Craig, P.C. 1984. Fish use of coastal waters of the Alaskan Beaufort Sea: A review. Trans. Am. Fish. Soc. 113:265-282.

Craig, P.C, W.B. Griffiths, S.R. Johnson, and D.M. Schell. 1984. Trophic dynamics in an arctic lagoon. Pp. 347-380 in The Alaskan Beaufort Sea: Ecosystems and Environment, P.W. Barnes, E. Reimnitz, and D.M. Schell, eds. New York: Academic Press.

Craig, J.D., K.W. Sherwood, and P.P. Johnson. 1985. Geologic Report for the Beaufort Sea Planning Area: Regional Geology, Petroleum Geology, Environmental Geology. OCS Study MMS 85-0111. Minerals Management Service, U.S. Department of the Interior, Anchorage, Alaska. Available as NTIS PB 86-218203.

Creighton, J.L. 1981. Public Involvement Manual. Cambridge, Mass.: Abt Books. 344 pp.

Cummings, R.G., and W.D. Schulze. 1978. Optimal investment strategy for boomtowns: A theoretical analysis. Am. Econ. Rev. 68(3):374-385.

Danielewicz, B.W., and D. Blanchet. 1988. Multi-year ice loads on Hans Island during 1980 and 1981. Pp. 465-484 in POAC 87: Proceedings of the International Conference on Port and Ocean Engineering Under Arctic Conditions, Vol. 1, W.M. Sackinger and M.O. Jeffries, eds. Geophysical Institute, University of Alaska, Fairbanks.

Davies, J.M., D.R. Bedborough, R.A.A. Blackman, J.M. Addy, J.F. Applebee, W.C. Grogan, J.G. Parker, and A. Whitehead. 1989. Environmental effect of oil-based mud drilling in the North Sea. Pp. 59-89 in Drilling Wastes: Proceedings of the 1988 International Conference on Drilling Waste, F.R. Engelhardt, J.P. Ray, and A.H. Gillam, eds. New York: Elsevier Applied Science.

Davis, J.M., and J.R. Pollock. 1992. With Prudhoe Bay in decline, what's next for Alaska? Oil Gas J. Special (Aug. 3).

Derksen, D.V., T.C. Rothe, and W.D. Eldridge. 1981. Use of Wetland Habitats by Birds in the National Petroleum Reserve: Alaska. USFWS Resource Publ. No. 141. U.S. Fish and Wildlife Service, U.S. Department of the Interior, Washington, D.C. 27 pp.

Divoky, G.J. 1976. The pelagic feeding habits of Ivory and Ross' gulls. Condor 78:85-90.

Divoky, G.J. 1978a. Identification, documentation and delineation of coastal migratory bird habitats in Alaska. I. Breeding bird use of barrier islands in the northern Chukchi and Beaufort seas. Partial Final Report. Environmental Assessment of the Alaskan Continental Shelf, Annual Reports 1:482-569.

Divoky, G.J. 1978b. The distribution, abundance and feeding ecology of birds associated with pack ice. Environmental Assessment of the Alaskan Continental Shelf, Quarterly Reports, April-June 251-252.

Divoky, G.J. 1983. The pelagic and nearshore birds of the Alaskan Beaufort Sea. National Oceanic and Atmospheric Administration/Outer

Continental Shelf Environmental Assessment Program (NOAA/ OCSEAP) Final Reports of Principal Investigators 23(1984):397-513.

Divoky, G.J. 1984. The pelagic and nearshore birds of the Alaskan Beaufort Sea: Biomass and trophics. Pp. 417-437 in The Alaska Beaufort Sea: Ecosystems and Environment, P.W. Barnes, D.M. Schell, and E. Reimnitz, eds. New York: Academic Press.

Divoky, G.J. 1987. The distribution and abundance of birds in the eastern Chukchi sea in late summer and early fall. Final Report. National Oceanic and Atmospheric Administration/Outer Continental Shelf Environmental Assessment Program (NOAA/OCSEAP). Arctic Environmental Information and Data Center, Anchorage, Alaska. 91 pp.

Divoky, G.J. 1991. The distribution and abundance of birds in the Bering Sea packs ice in spring and early summer. Final Report. National Oceanic and Atmospheric Administration/Outer Continental Shelf Environmental Assessment Program (NOAA/OCSEAP). Arctic Environmental Information and Data Center, Anchorage, Alaska.

DOE (U.S. Department of Energy). 1993. Petroleum Supply Annual 1992, Vol. 1. DOE/Energy Information Administration (DOE/EIA)-0340(92)/1. U.S. Department of Energy, Washington, D.C.

Drielsma, J.H. 1984. The Influence of Forest-Based Industries on Rural Communities. Ph.D. Dissertation. Department of Sociology, Yale University, New Haven, Conn.

Dugger, J.A. 1984. Arctic oil and gas: Policy perspectives. Pp. 19-38 in United States Arctic Interests: The 1980s and 1990s, W.E. Westermeyer and K.M. Shusterich, eds. New York: Springer-Verlag.

Dunbar, M.J. 1981. Physical causes and biological significance of polynyas and other open water in sea ice. Pp. 29-43 in Polynyas in the Canadian Arctic, I. Stirling and H. Cleator, eds. Canadian Wildlife Service Occasional Paper No. 45. Available from Canadian Government Publication Centre, Quebec., Cat. No. CW 69-1/45E.

Dunbar, M.J. 1985. The Arctic marine ecosystem. Pp. 1-35 in Petroleum Effects in the Arctic Environment, F.R. Engelhardt, ed. New York: Elsevier Applied Science. 272 pp.

Dunton, K.H. 1992. Arctic biogeography: The paradox of the marine benthic fauna and flora. Trends Ecol. Evol. 7:183-189.

Dykins, J.E. 1971. Ice Engineering: Material Properties of Saline Ice for a Limited Range of Conditions. Tech. Rep. R-720. Naval Civil Engineering Laboratory, Port Hueneme, Calif.

Elo, I.T., and C.L. Beale. 1985. Natural Resources and Rural Poverty: An Overview. National Center for Food and Agricultural Policy. Resources for the Future, Washington, D.C.

Elphick, C.S., and G.L. Hunt, Jr. 1993. Variations in the distributions of marine birds with watermass in the northern Bering Sea. Condor 95:33-44.

Engelhardt, F.R. 1985a. Petroleum Effects in the Arctic Environment. New York: Elsevier Applied Science. 272 pp.

Engelhardt, F.R. 1985b. Environmental issues in the Arctic. In POAC 85: Proceedings of the International Conference on Port and Ocean Engineering Under Arctic Conditions, Sept. 1985, Narssarssuaq, Greenland. Horsholm, Denmark: Danish Hydraulic Institute.

Eppley, Z.A., and G.L. Hunt, Jr. 1984. Pelagic distribution of marine birds on the central Bering Sea shelf and analysis of risk for the Navarin Basin. National Oceanic and Atmospheric Administration/Outer Continental Shelf Environmental Assessment Program (NOAA/OCSEAP) Final Reports of Principal Investigators 30(1985):357-526.

Evans, D.D. 1991. In situ burning as an oil spill response technology. Pp. 195-200 in Technology Assessment and Research Program for Offshore Minerals Operations. OCS Study MMS 91-0057. Minerals Management Service, U.S. Department of the Interior, Herndon, Va.

Fadely, B.S., J.F. Piatt, S.A. Hatch, and D.G. Roseneau. 1989. Populations, Productivity, and Feeding Habits of Seabirds of Cape Thompson, Alaska. OCS Study MMS 89-0014. Alaska Fish and Wildlife Research Center, U.S. Fish and Wildlife Service, U.S. Department of the Interior, Anchorage, Alaska. Available as NTIS PB 90-161639. 429 pp.

Fay, F.H. 1974. The role of ice in the ecology of marine mammals in the Bering Sea. Pp. 383-389 in Oceanography of the Bering Sea: With Emphasis on Renewable Resources, D.W. Hood and E.J. Kelley, eds. Occasional Publ. Ser. No 2. Institute of Marine Science, University of Alaska, Fairbanks.

Fay, F.H. 1982. Ecology and Biology of the Pacific Walrus, *Odobenus rosmarus divergens* (Illiger, 1815). North American Fauna No. 74. U.S. Fish and Wildlife Service, U.S. Department of the Interior, Washington, D.C. 279 pp.

Feder, H.M. A.S. Naidu, M.J. Hameedi, S.C. Jewett, and W.R. Johnson. 1989. The Chukchi Sea Continental Shelf: Benthos—Environmental

Interactions. National Oceanic and Atmospheric Administration/Outer Continental Shelf Environmental Assessment Program (NOAA/ OCSEAP) Final Reports of Principal Investigators 68:25-311.

Fisher, W.L., N. Tyler, C.L. Ruthven, T. Burchfield, and J.F. Pautz. 1992. An Assessment of the Oil Resource Base of the United States. DOE/BC-93/1/SP. U.S. Department of Energy, Washington, D.C.

Fitzgerald, T.A. 1980. Giant Field Discoveries 1968-1978: An Overview. AAPG Memoir '30. American Association of Petroleum Geologists, Tulsa, Okla.

Fjeld, P.E., G.W. Gabrielson, and J.B. Orbeck. 1988. Noise from helicopters and its affect on a colony of Brünnich's guillemots (Uria lowvia) on Svalbard. North Polarinstitutt Rapportserie 41:115-153.

Ford, R.G., J.A. Wiens, D. Heinemann, and G.L. Hunt, Jr. 1982. Modelling the sensitivity of colonially breeding marine birds to oil spills: Guillemot and kittiwake populations on the Pribilof Islands, Bering Sea. J. Appl. Ecol. 19:1-31.

Freudenburg, W.R. 1988. Perceived risk, real risk: Social science and the art of probabilistic risk assessment. Science 242:44-49.

Freudenburg, W.R., and R. Gramling. 1992. Community impacts of technological change: Toward a longitudinal perspective. Social Forces 70:937-957.

Freudenburg, W.R., and R.E. Jones. 1992. Criminal behavior and rapid community growth: Examining the evidence. Rural Sociol. 56(Jan. 4):619-645.

Frost, K.J., and L.F. Lowry. 1981. Ringed, Baikal and Caspian seals—Phoca hispida, Phoca sibirica and Phoca caspica. Pp. 29-53 in Handbook of Marine Mammals: Seals, Vol. 2, S.H. Ridgway and R.J. Harrison, eds. New York: Academic Press.

Frost, K.J., and L.F. Lowry. 1984. Trophic relationships of vertebrate consumers in the Alaskan Beaufort Sea. Pp. 381-401 in The Alaskan Beaufort Sea: Ecosystems and Environments, P.W. Barnes, D.M. Schell, and E. Reimnitz, eds. New York: Academic Press.

Frost, K.J., and L.F. Lowry. 1990. Distribution, abundance, and movements of beluga whales, Delphinapterus leucas, in coastal waters of western Alaska. Pp. 39-57 in Canadian Bulletin of Fisheries and Aquatic Sciences 224: Advances in Research on the Beluga Whale, Delphinapterus leucas, T.G. Smith, D.J. St. Aubin, and J.R. Geraci, eds. Department of Fisheries and Oceans, Ottawa, Ontario.

Frost, K.J., L.F. Lowry, J.R. Gilbert, and J.J. Burns. 1988. Ringed seal monitoring: Relationships of distribution and abundance to habitat attributes and industrial activities. National Oceanic and Atmospheric Administration/Outer Continental Shelf Environmental Assessment Program (NOAA/OCSEAP) Final Reports of Principal Investigators 61(1989):345-445.

Frost, K.J., L.F. Lowry, and G. Carroll. 1993. Beluga whale and spotted seal use of a coastal lagoon system in the northeastern Chukchi Sea. Arctic 46(1).

Frost, K.J., L.F. Lowry, E. Sinclair, J. Ver Hoef, and D.C. McAllister. In press. Impacts on distribution, abundance, and productivity of harbor seals. In Impacts of the *Exxon Valdez* Oil Spill on Marine Mammals. San Diego, Calif.: Academic Press.

GAO (U.S. General Accounting Office). 1993. Trans-Alaska Pipeline: Projections of Long-Term Viability Are Uncertain. GAO/RCED-93-69. U.S. General Accounting Office, Washington, D.C.

Garner, G.W., S.T. Knick, and D.C. Douglas. 1990. Seasonal movements of adult female polar bears in the Bering and Chukchi seas. Int. Conf. Bear Res. Manage. 8:219-226.

GESAMP (Group of Experts on the Scientific Aspects of Marine Pollution). 1992. Review of Potentially Harmful Substances: Oil, and Other Hydrocarbons Including Used Lubricating Oils, Oil Spill Dispersants and Chemicals Used in Offshore Exploration and Exploitation. International Maritime Organization, London, U.K.

Giammona, C.P., F.R. Engelhardt, and J. Osborne. 1992. Applied uses, needs and issues concerning oil spill trajectory models. Pp. 186-189 in Ocean Model Workshop Proceedings. Can. Tech. Rep. Hydrogr. Ocean Sci. Rep. No. 140. Department of Fisheries and Oceans, Bedford Institute of Oceanography, Dartmouth, Nova Scotia.

Gilbert, J., G. Fedoseev, D. Seagars, E. Razlivalov, and A. Lachugin. 1992. Aerial Census of Pacific Walrus, 1990. USFWS Admin. Rep. R7/MMM 92-1. U.S. Fish and Wildlife Service, U.S. Department of the Interior, Washington, D.C. 33 pp.

Goodwin, C.R., C.R. Finley, and L.M. Howard, eds. 1985. Ice Scour Bibliography. Environmental Studies Revolving Funds Report No. 010. Arctic Institute of North America, University of Calgary, Calgary, Alberta. 99 pp.

Gould, P.J., D.J. Forsell, and C.J. Lensink. 1982. Pelagic Distribution and Abundance of Seabirds in the Gulf of Alaska and Eastern Bering Sea. U.S. Fish and Wildlife Service, U.S. Department of the Interior, Washington, D.C. 294 pp.

Grantz, A., L. Johnson, and J.F. Sweeney, eds. 1990. Geology of North America: The Arctic Ocean Region, Vol. L. Geological Society of America, Boulder, Colo. 644 pp.

Gregory, R., H. Kunreuther, D. Easterling, and K. Richards. 1991. Incentives policies to site hazardous waste facilities. Risk Anal. 11:667-675.

Gulliford, A. 1989. Boomtown Blues: Colorado Oil Shale, 1885-1985. Niwot, Colo.: University Press of Colorado.

Guzman, J.R., and M.T. Myers. 1987. Ecology and behavior of southern hemisphere shearwater (*Genus puffinus*) when over the outer continental shelf, the Gulf of Alaska and Bering Sea during the northern summer (1975-1976). National Oceanic and Atmospheric Administration/Outer Continental Shelf Environmental Assessment Program (NOAA/OCSEAP) Final Reports of Principal Investigators 54:571-682.

Haimila, N.E., C.E. Kirschner, W.W. Nassichuk, G. Ulmichek, and R.M. Procter. 1990. Sedimentary basins and petroleum resource potential of the Arctic Ocean region. Pp. 503-538 in The Geology of North America: The Arctic Ocean Region, Vol. L, A. Grantz, L. Johnson, and J.F. Sweeney, eds. Geological Society of America, Boulder, Colo.

Hance, B.J., C. Chess, and P.M. Sandman. 1988. Improving Dialogue with Communities: A Risk Communication Manual for Government. New Jersey Department of Environmental Protection, New Brunswick, N.J. 83 pp.

Haney, J.C. 1991. Influence of pyconocline topography and water-column structure on marine distributions of alcids (*Aves: Alcidae*) in Anadyr Strait, Northern Bering Sea, Alaska. Mar. Biol. 110:419-435.

Hansell, D.A., J.J. Goering, J.J. Walsh, C.P. McRoy, L.K. Coachman, and T.E. Whitledge. 1989. Summer phytoplankton production and transport along the shelf break in the Bering Sea. Continent. Shelf Res. 9:1085-1104.

Harrison, C.S. 1979. The association of marine birds and feeding grey whales. Condor 81:93-95.

Hills, S. 1992. The Effect of Spatial and Temporal Variability on Population Assessment of Pacific Walruses. Ph.D. Dissertation. University of Maine, Orono, Maine. 208 pp.

Hertz, H. 1884. Uber das gleichgewicht schimmender elasticher platten. Wiedmanns Ann. Phys. Chem. 22:449-455.

Hickman, H.L. 1988. It's a national mental illness. Waste Age (March):197-198.

Holmes, W.N., and J. Cronshaw. 1977. Biological effects of petroleum on marine birds. Pp. 359-398 in Effects of Petroleum and Arctic and Subarctic Marine Environments and Organisms: Biological Effects, Vol. 2. New York: Academic Press.

Hubbard, R.J., S.P. Edrich, and R.P. Rattey. 1987. Geologic evolution and hydrocarbon habitat of the arctic Alaska microplate. Mar. Petrol. Geol. 4:2-34.

Hum, S. 1977. The development and use of resource sensitivity maps for oil spill countermeasures. Proceedings of 1977 Oil Spill Conference. Am. Petrol. Inst. Publ. 4284:105-110.

Humphrey, B., G. Sergy, and E.H. Owens. 1990. Stranded oil persistence in cold climates. Pp. 401-410 in Proceedings of the 13th Arctic and Marine Oil Spill Program Technical Seminar. Technology Development Branch, Environment Canada, Ottawa, Ontario.

Hunt, G.L., Jr. 1987. Offshore oil development and seabirds: The present status of knowledge and long-term research needs. Pp. 539-586 in Long-Term Environmental Effects of Offshore Oil and Gas Development, D.F. Boesch and N.N. Rabalais, eds. New York: Elsevier Applied Science.

Hunt, G.L., Jr. 1991. Marine birds and ice-influenced environments of polar oceans. J. Mar. Syst. 2:233-240.

Hunt, G.L., Jr., and N.M. Harrison. 1990. Foraging habitat and prey taken by least auklets at King Island, Alaska. Mar. Ecol. Prog. Ser. 65:141-150.

Hunt, G.L., Jr., B. Burgeson, and G.A. Sanger. 1981a. Feeding ecology of seabirds of the eastern Bering Sea. Pp. 629-647 in The Eastern Bering Sea Shelf: Oceanography and Resources, Vol. 2, D.W. Hood and J.A. Calder, eds. National Oceanic and Atmospheric Administration, Washington, D.C.

Hunt, G.L., Jr., P.J. Gould, D.J. Forsell, and H. Peterson, Jr. 1981b. Pelagic distribution of marine birds in the eastern Bering Sea. Pp. 689-718 in The Eastern Bering Sea Shelf: Oceanography and Resources, Vol. 2, D.W. Hood and J.A. Calder, eds. National Oceanic and Atmospheric Administration, Washington, D.C.

Hunt, G.L., Jr., N.M. Harrison, W.M. Hamner, and B.S. Obst. 1988. Observations of a mixed-species flock of birds foraging on euphausiids near St. Matthew Island, Bering Sea. Auk 105:345-349.

Hunt, G.L., Jr., N.M. Harrison, and R.T. Cooney. 1990. The influence of hydrographic structure and prey abundance on foraging of least auklets. Stud. Avian Biol. 14:7-22.

Impact Assessment, Inc. 1990. Economic, Social, and Psychological Impact Assessment of the *Exxon Valdez* Oil Spill. Report prepared for the Oiled Mayors Subcommittee, Alaska Conference of Mayors. Impact Assessment, Inc., La Jolla, Calif.

Interagency Coordinating Committee on Oil Pollution Research. 1992. Oil Pollution Research and Technology Plan. U.S. Coast Guard, Washington, D.C.

IWC (International Whaling Commission). 1989. Report of the Scientific Committee of the International Whaling Commission. Rep. Int. Whal. Comm. 39:33-157.

IWC (International Whaling Commission). 1990. Report of the Special Meeting of the Scientific Committee on the Assessment of Gray Whales. Rep. Int. Whal. Comm. 40.

Jamison, P., S. Zegura, and F. Milan, eds. 1978. Eskimo of Northwestern Alaska: A Biological Perspective. New York: Van Nostrand Reinhold.

Jeffries, M.O. 1992. Arctic ice shelves and ice islands: Origin, growth and disintegration, physical characteristics, structural-stratigraphic variability, and dynamics. Rev. Geophys. 30:245-267.

Jeffries, M.O., and W.M. Sackinger. 1990. Near-real-time, synthetic aperture radar detection of a calving event at the Milne Ice Shelf, N.W.T. Pp. 320-331 in Ice Technology for Polar Operations, T.K.S. Murthy, J.G. Paren, W.M. Sackinger, and P. Wadhams, eds. Proceedings of the Second International Conference on Ice Technology. Cambridge, Mass.: Computational Mechanics Publications.

Johnson, S.R. 1983. Birds and marine mammals. In Environmental Characterization and Biological Use of Lagoons in the Eastern Bering Sea. National Oceanic and Atmospheric Administration/Outer Continental Shelf Environmental Assessment Program (NOAA/OCSEAP) Final Reports of Principal Investigators 24(1984):265-324.

Johnson, S.R. 1984a. Prey selection by Oldsquaws (*Clangula hyemalis L.*) in a Beaufort Sea lagoon. Pp. 12-19 in Marine Birds: Their Feeding

Ecology and Commercial Fisheries Relationships. Canadian Wildlife Service Special Publication, Ottawa, Ontario.

Johnson, S.R. 1984b. Habitat Use and Behavior of Nesting Common Eiders and Molting Oldsquaws at Thetis Island, Alaska During a Period of Industrial Activity. Report prepared by LGL Alaska Research Associates for Sohio Alaska Petroleum Co., Anchorage, Alaska.

Johnson, S.R., and W.J. Gazey. 1992. Design and testing of a monitoring program for Beaufort Sea waterfowl and marine birds. Report prepared by LGL Alaska Research Associates. OCS Study MMS 92-0060. Minerals Management Service, U.S. Department of the Interior, Herndon, Va. 114 pp.

Johnson, S.R., and D.R. Herter. 1989. The Birds of the Beaufort Sea. BP Exploration (Alaska), Inc., Anchorage, Alaska. 372 pp.

Johnson, S.R., and W.J. Richardson. 1981. Beaufort Sea barrier island-lagoon ecological process studies: Part 3: Birds. Final Report, Simpson Lagoon. Environmental Assessment of the Alaskan Continental Shelf, Final Reports, Biological Studies 7:109-383. Available as NTIS PB 82-192113/AS.

Johnson, S.R., P.G. Connors, G.J. Divoky, R. Meehan, and D.W. Norton. 1987. Coastal and marine birds. Pp. 131-145 in The Diapir Field Environment and Possible Consequences of Planned Offshore Oil and Gas Development, P.R. Becker, ed. National Oceanic and Atmospheric Administration/Outer Continental Shelf Environmental Assessment Program (NOAA/OCSEAP) Proceedings of a Synthesis Meeting. OCS Study MMS 85-0092. Minerals Management Service, U.S. Department of the Interior, Anchorage, Alaska.

Johnson, S.R., D.A. Wiggins, and P.F. Warnwright. 1992. II: Marine birds. Pp. 57-510 in Use of Kasegaluk Lagoon, Chukchi Sea, Alaska, by Marine Birds and Mammals. Report prepared by the LGL Alaska Research Associates and the Alaska Department of Fish and Game. MMS 92-0028. Minerals Management Service, U.S. Department of the Interior, Anchorage, Alaska.

Jones, M.L., S.L. Swartz, and S.L. Leatherwood. 1984. The Gray Whale: *Eschrichtius robustus*. Orlando, Fla: Academic Press. 600 pp.

Jorgenson, J.G. 1990. Oil Age Eskimos. Los Angeles: University of California Press.

Kinder, T.H., and J.D. Schumacher. 1981a. Circulation over the continental shelf of the southeastern Bering Sea. Pp. 53-75 in The

Eastern Bering Sea Shelf: Oceanography and Resources, Vol. 1, D.W. Hood and J.A. Calder, eds. National Oceanic and Atmospheric Administration, Washington, D.C.

Kinder, T.H., and J.D. Schumacher. 1981b. Hydrographic structure over the continental shelf of the southeastern Bering Sea. Pp. 31-52 in The Eastern Bering Sea Shelf: Oceanography and Resources, Vol. 1, D.W. Hood and J. A. Calder, eds. National Oceanic and Atmospheric Administration, Washington, D.C.

Kirstein, B.E., and R.T. Redding. 1987. Ocean-ice oil-weathering computer program user's manual. National Oceanic and Atmospheric Administration/Outer Continental Shelf Environmental Assessment Program (NOAA/OCSEAP) Final Reports of Principal Investigators 59(1988):1-145.

Kovacs, A. 1977. Sea ice thickness profiling and under-ice oil entrapment. Pp. 547-554 in 1977 Proceedings of the Ninth Annual Offshore Technology Conference. Offshore Technology Conference, Houston, Tex.

Kovacs, A., and R.M. Morey. 1986. Electromagnetic measurements of multi-year sea ice using impulse radar. Cold Regions Sci. Technol. 12:6793.

Kovacs, A., and J.S. Holladay. 1990. Sea ice thickness measurement using a small airborne electromagnetic sounding system. Geophysics 55(10):1327-1337.

Kowalik, Z. 1984. Storm surges in the Beaufort and Chukchi seas. J. Geophys. Res. 89:10570-10578.

Kowalik, Z., and J.B. Matthews. 1982. The M_2 tide in the Beaufort and Chukchi seas. J. Phys. Oceanogr. 12:743-746.

Kowalik, Z., and J.B. Matthews. 1983. Numerical study of the water movement driven by brine rejection from nearshore arctic ice. J. Geophys. Res. 88:2953-2958.

Kozo, T.L. 1984. Mesoscale wind phenomena along the Alaskan Beaufort Sea coast. Pp. 23-45 in The Alaskan Beaufort Sea: Ecosystems and Environment, P.W. Barnes, D.M. Schell, and E. Reimnitz, eds. Orlando, Fla.: Academic Press.

Krannich, R.S., and A.E. Luloff. 1991. Problems of resource dependency in U.S. rural communities. Prog. Rural Policy Plann. 1:5-18.

Kruse, J., J. Kleinfeld, and R. Travis. 1981. Energy Development and the North Slope Inupiat: Quantitative Analysis of Social and Economic

Change. Man in the Arctic Monograph No. 1. Institute of Social and Economic Research, University of Alaska, Anchorage.

Kruse, J.A., J. Kleinfeld, and R. Travis. 1982. Energy Development on Alaska's North Slope: Effects on the Inupiat Population. Hum. Organ. 41:97-106.

Kumar, N. 1992. Arctic Exploration: Challenges and Opportunities. Report presented to the National Research Council's Committee to Review Alaskan Outer Continental Shelf Environmental Information, May 6, Irvine, Calif.

Kvenvolden, K.A., and A. Grantz. 1990. Gas hydrates of the Arctic Ocean region. Pp. 539-549 in Geology of North America: The Arctic Ocean Region, Vol. L, A. Grantz, L. Johnson, and J.F. Sweeney, eds. Geological Society of America, Boulder, Colo.

Kvitek, R.G., and J.S. Oliver. 1986. Side-scan sonar estimates of the utilization of gray whale feeding grounds along Vancouver Island, Canada. Continent. Shelf Res. 6:639-654.

Lake, R.A., and E.L. Lewis. 1970. Salt rejection by sea ice during growth. J. Geophys. Res. 75:583-597.

Lee, J., T.D. Ralston, and D.H. Petrie. 1986. Full-thickness sea ice strength tests. Pp. 293-306 in Proceedings of the IAHR (International Association for Hydraulic Research) Ice Symposium 1986, Iowa City, Iowa. Iowa City: University of Iowa Press.

Lentfer, J.W., ed. 1988. Selected Marine Mammals of Alaska: Species Accounts with Research and Management Recommendations. Marine Mammal Commission, U.S. Department of the Interior, Washington, D.C. Available as NTIS PB 88-178462. 275 pp.

Lentfer, J.W., convener. 1990. Workshop on Measures to Assess and Mitigate the Adverse Effects of Arctic Oil and Gas Activities on Polar Bears, Jan. 24-25, 1989, Anchorage, Alaska. Prepared for the Marine Mammal Commission, U.S. Department of the Interior, Washington, D.C. Available as NTIS PB 91-127241. 39 pp.

Lewis, E.L. 1976. Oil in sea ice. Pp. 229-260 in Science in Alaska. II: Resource Development Processes and Problems. American Association for the Advancement of Science, Washington, D.C.

Lewis, J.W., and R.Y. Edwards. 1970. Methods for predicting icebreaking and ice resistance characteristics of icebreakers. Soc. Nav. Archit. Mar. Eng. Trans. 78.

LGL Alaska Research Associates, Inc. 1992. The 1991 Endicott Development Fish Monitoring Program, Vols. 1-2. Prepared for BP Exploration (Alaska) Inc. and North Slope Borough. LGL Alaska Research Associates, Inc., Anchorage, Alaska.

Lindstedt-Siva, J. 1977. Oil spill response planning for biologically sensitive areas. Proceedings of 1977 Oil Spill Conference. Am. Petrol. Inst. Publ. 4284:111-114.

Lindstedt-Siva, J. 1979a. Ecological impacts of oil spill cleanup: Are they significant? Proceedings of 1979 Oil Spill Conference. Am. Petrol. Inst. Publ. 4308:521-524.

Lindstedt-Siva, J. 1979b. Why clean up oil spills: Another look. Spill Technol. Newslett. 4(1):15-16.

Lindstedt-Siva, J., B.J. Bacca, and C.D. Getter. 1983. The MIRG environmental element: An oil spill response planning tool for the Gulf of Mexico. Proceedings of 1983 Oil Spill Conference. Am. Petrol. Inst. Publ. 4356:175-181.

Lindstedt-Siva, J. 1984. Oil spill response and ecological impacts—15 years beyond Santa Barbara. Mar. Technol. Soc. J. 18(3):43-50.

Lønne, O.J., and G.W. Gabrielsen. 1992. Summer diet of seabirds feeding in sea-ice covered waters near Svolbard. Polar Biol. 12:685-692.

Lowry, L.F. 1993. Foods and feeding ecology. Pp. 201-238 in The Bowhead Whale, Society for Marine Mammalogy Special Publ. No. 2, J.J. Burns, J.J. Montague, and C.J. Cowles, eds. Lawrence, Kans.: Allen Press.

Lowry, L.F., and K.J. Frost. 1981. Distribution, growth, and foods of arctic cod (Boreogadus saida) in the Bering, Chukchi, and Beaufort seas. Can. Field-Natural. 95:186-191.

Lowry, L.F., K.J. Frost, and J.J. Burns. 1980a. Variability in the diet of ringed seals, Phoca hispida, in Alaska. Can. J. Fish. Aquat. Sci. 37:2254-2261.

Lowry, L.F., K.J. Frost, and J.J. Burns. 1980b. Feeding of bearded seals in the Bering and Chukchi seas and trophic interaction with Pacific walruses. Arctic 33:330-342.

Lowry, L.F., J.J. Burns, and K.J. Frost. 1989. Recent harvests of belukha whales, Delphinapterus leucas, in western and northern Alaska and their potential impact on provisional management stocks. Rep. Int. Whal. Comm. 39:335-339.

Luton, H. 1985. Effects of Renewable Resource Harvest Disruptions on

Socioeconomic and Sociocultural Systems. Alaska OCS Social and Economic Studies Program. Tech. Rep. No. 91. Alaska OCS Region, Minerals Management Service, U.S. Department of the Interior, Anchorage, Alaska.

Macdonald, R.W., E.C. Carmack, F.A. McLaughlin, M. O'Brian, and J.E. Papadakis. 1989. Composition and modification of water masses in the Mackenzie shelf/estuary. J. Geophys. Res. 94:18057-18070.

Mackay, D. 1985. The physical and chemical fate of spilled oil. Pp. 37-61 in Petroleum Effects in the Arctic Environment, F.R. Engelhardt, ed. New York: Elsevier Applied Science.

Makarov, S.O. 1901. The Ermak in the ice fields: A description of the construction and voyages of the ice breaker Ermak and a summary of scientific results from the voyages. St. Petersburg, Tipografiia Sankt-petersburgskogo aktsionnogo obshchestvo pechatnogo dela v Rosii E. Evdokimov.

Malins, D.C., ed. 1977. Effects of Petroleum on Arctic and Subarctic Marine Environments and Organisms: Nature and Fate of Petroleum, Vol. 1. New York: Academic Press.

Marshall, P.G. 1989. Not in my backyard! Congressional Quarterly's Editorial Res. Rep. (June 9):306-318.

McBeath, G. 1981. North Slope Borough Government and Policymaking. Institute of Social and Economic Research, University of Alaska, Fairbanks.

Melling, H., and E.L. Lewis. 1982. Shelf drainage flows in the Beaufort Sea and their effect on the Arctic Ocean pycnocline. Deep-Sea Res. 29:967-985.

Mendenhall, V.M., ed. 1993. Monitoring Seabird Populations in Areas of Oil and Gas Development on the Alaskan Continental Shelf: Monitoring Populations and Productivity of Seabirds at Cape Pierce, Bluff, and Cape Thompson, Alaska, 1990. OCS Study MMS 592-0047. Minerals Management Service, U.S. Department of the Interior, Anchorage, Alaska.

Menzie, C.A. 1982. The environmental implications of offshore oil and gas activities. Environ. Sci. Technol. 16:454A-472A.

Michel, J., M.D. Hayes, and P.J. Brown. 1978. Application of an oil spill vulnerability index to the shoreline of lower Cook Inlet, Alaska. Environ. Geol. 2(2):107-117.

MMS (Minerals Management Service). 1987a. Leasing Energy Resources on the Outer Continental Shelf. Offshore Minerals Management Program, Minerals Management Service, U.S. Department of the Interior, Herndon, Va.

MMS (Minerals Management Service). 1987b. Five-Year Outer Continental Shelf Oil and Gas Leasing Program, Proposed Mid-1987 to Mid-1992. Final Environmental Impact Statement. OCS EIS/EA, MMS 86-0127. Minerals Management Service, U.S. Department of the Interior, Herndon, Va. 3 vols.

MMS (Minerals Management Service) 1987c. Rigs-to-Reefs: The Use of Obsolete Petroleum Structures as Artificial Reefs. OCS Report MMS 87-0015. Gulf of Mexico Region, Minerals Management Service, U.S. Department of the Interior, New Orleans, La. 17 pp.

MMS (Minerals Management Service). 1987d. Geologic Report for the Chukchi Sea Planning Area, Alaska. OCS Report MMS 87-0046. Alaska Region, Minerals Management Service, U.S. Department of the Interior, Anchorage, Alaska. Available as NTIS PB 87-222501.

MMS (Minerals Management Service). 1989. Estimates of Undiscovered Oil and Gas Resources for the Outer Continental Shelf as of January 1987. OCS Report MMS 89-0090. Minerals Management Service, U.S. Department of the Interior, Herndon, Va. 114 pp.

MMS (Minerals Management Service). 1990. Beaufort Sea Planning Area Oil and Gas Lease Sale 124. Final Environmental Impact Statement. OCS EIS/EA, MMS 90-0063. Alaska Region, Minerals Management Service, U.S. Department of the Interior, Anchorage, Alaska.

MMS (Minerals Management Service). 1991a. OCS National Compendium: Outer Continental Shelf Oil and Gas Information through October 1990. OCS Information Report, MMS 91-0032. Minerals Management Service, U.S. Department of the Interior, Herndon, Va.

MMS (Minerals Management Service). 1991b. Chukchi Sea Oil and Gas Lease Sale 126. Final Environmental Impact Statement. OCS EIS/EA, MMS 90-0095. Alaska Region, Minerals Management Service, U.S. Department of the Interior, Anchorage, Alaska. 2 vols.

MMS (Minerals Management Service). 1991c. Navarin Basin Oil and Gas Lease Sale 107. Final Environmental Impact Statement. OCS EIS/EA, MMS 91-0008. Alaska Region, Minerals Management Service, U.S. Department of the Interior, Anchorage, Alaska.

MMS (Minerals Management Service). 1992a. Oil and Gas Leasing/Production Program: Annual Report/FY 1991. OCS Report MMS

92-0059. Minerals Management Service, U.S. Department of the Interior, Herndon, Va.

MMS (Minerals Management Service). 1992b. Federal Offshore Statistics: 1991. Leasing, Exploration, Production, and Revenues as of December 31, 1991. MMS 92-0056. Minerals Management Service, U.S. Department of the Interior, Herndon, Va.

MMS (Minerals Management Service). 1992c. Comprehensive Program, 1992-1997. Proposed Final. OCS Natural Gas and Oil Resource Management, Minerals Management Service, U.S. Department of the Interior, Herndon, Va.

MMS (Minerals Management Service). 1992d. ESDS (Environmental Studies Database System). Minerals Management Service, U.S. Department of the Interior, Herndon, Va.

Moir, M.E., and D.C. Yetman. 1993. The detection of oil under ice by pulsed ultraviolet fluorescence. Pp. 521-523 in Proceedings of the International Oil Spill Conference, March, Tampa, Fla. American Petroleum Institute, Washington, D.C.

Montemurro, M.P., and J.F. Sykes. 1989. Evaluation of constitutive laws for sea ice with application to Adam's Island. Pp. 367-376 in POAC 89: Proceedings of the 10th International Conference on Port and Ocean Engineering Under Arctic Conditions, June 12-16, 1989, Lulea, Sweden, K.B.E. Axelsson and L. A. Fransson, eds. Lulea, Sweden: Lulea University of Technology..

Moore, S.E., and R.R. Reeves. 1993. Distribution and movement. Pp. 313-386 in The Bowhead Whale, Society for Marine Mammalogy Special Publ. No. 2, J.J. Burns, J.J. Montague, and C.J. Cowles, eds. Lawrence, Kans.: Allen Press.

Moran, E.R. 1982. Human Adaptability: An Introduction to Ecological Anthropology. Boulder, Colo.: Westview Press. 404 pp.

Muench, R.D. 1990. Mesoscale phenomena in the Polar oceans. Pp. 223-285 in Polar Oceanography. Part A: Physical Science, Walker O. Smith, Jr., ed. New York: Academic Press.

Muench, R.D., and K. Ahlnas. 1976. Ice movement and distribution in the Bering Sea from March to June 1974. J. Geophys. Res. 81:4467-4476.

Neff, J.M., N.N. Rabalais, and D.F. Boesch. 1987. Offshore oil and gas development activities potentially causing long-term environmental effects. Pp. 149-173 in Long-Term Environmental Effects of Offshore

Oil and Gas Development, D.F. Boesch and N.N. Rabalais, eds. New York: Elsevier Applied Science.

Neill, C.R. 1976. Dynamic ice forces on piers and piles: An assessment of design guidelines in the light of recent research. Can. J. Civ. Eng. 3:305-341.

Nelson, R.K. 1969. Hunters of the Northern Ice. Chicago, Ill.: University of Chicago Press.

Nero and Associates, Inc. 1987. Seabird Oil Toxicity Study. Final Report. OCS Study MMS 87-0005. Pacific Region, Minerals Management Service, U.S. Department of the Interior, Camarillo, Calif. Available as NTIS PB 90-161787.

Nettleship, D.N., and P.G.H. Evans. 1985. Distributions and status of the Atlantic Alcidae. Pp. 53-154 in The Atlantic Alcidae: The Evolution, Distribution, and Biology of the Auks Inhabiting the Atlantic Ocean and Adjacent Water Areas, D.N. Nettleship, T.R. Birkhead, and J. Bedard, eds. Orlando, Fla.: Academic Press.

Nevel, D.E., R.E. Perham, and G.B. Hogue. 1977. Ice Forces on Vertical Piles. CRREL Rep. 77-10. Cold Regions Research and Engineering Laboratory, U.S. Army Corps of Engineers, Hanover, N.H.

NFAC (National Foreign Assessment Center). 1978. Polar Regions Atlas. Central Intelligence Agency, McLean, Va. 66 pp.

Niebauer, H.J. 1983. Multiyear sea ice variability in the eastern Bering Sea: An update. J. Geophys. Res. 88:2733-2742.

Niebauer, H.J. 1988. Effects of El Niño-southern oscillation and North Pacific weather patterns on interannual variability in the subarctic Bering Sea. J. Geophys. Res. 93:5051-5068.

Niebauer, H.J., and D.M. Schell. 1993. Physical environment of the Bering Sea population. Pp. 23-43 in The Bowhead Whale, Society for Marine Mammalogy Special Publ. No. 2, J.J. Burns, J.J. Montague, and C.J. Cowles, eds. Lawrence, Kan.: Allen Press.

NRC (National Research Council). 1975. Petroleum in the Marine Environment. Washington, D.C.: National Academy Press.

NRC (National Research Council). 1978. OCS Oil and Gas: An Assessment of the Department of the Interior Environmental Studies Program. Washington, D.C.: National Academy of Sciences.

NRC (National Research Council). 1983a. Drilling Discharges in the Marine Environment. Washington, D.C.: National Academy Press. 180 pp.

NRC (National Research Council). 1983b. Understanding the Arctic Sea Floor for Engineering Purposes. Washington, D.C.: National Academy Press.

NRC (National Research Council). 1983c. Navy Long Range Deep Ocean Technology. Washington, D.C.: National Academy Press.

NRC (National Research Council). 1985. Oil in the Sea. Inputs, Fates, and Effects. Washington, D.C.: National Academy Press.

NRC (National Research Council). 1989a. The Adequacy of Environmental Information for Outer Continental Shelf Oil and Gas Decisions: Florida and California. Washington, D.C.: National Academy Press.

NRC (National Research Council). 1989b. Using Oil Spill Dispersants on the Sea. Washington, D.C.: National Academy Press. 335 pp.

NRC (National Research Council). 1989c. Evaluation of the Hydrocarbon Resource Estimates for the Offshore Areas of Northern and Southern California and Florida South of 26° Latitude. Washington, D.C.: National Academy Press.

NRC (National Research Council). 1990a. Assessment of the U.S. Outer Continental Shelf Environmental Studies Program. I. Physical Oceanography. Washington, D.C.: National Academy Press.

NRC (National Research Council). 1990b. Adequacy of the Data Base for Hydrocarbon Estimates of the Georges Bank Area of the North Atlantic Outer Continental Shelf. Washington, D.C.: National Academy Press.

NRC (National Research Council). 1991a. The Adequacy of Environmental Information for Outer Continental Shelf Oil and Gas Decisions: Georges Bank. Washington, D.C.: National Academy Press.

NRC (National Research Council). 1991b. Undiscovered Oil and Gas Resources. An Evaluation of the Department of the Interior's 1989 Assessment Procedures. Washington, D.C.: National Academy Press.

NRC (National Research Council). 1991c. Tanker Spills: Prevention by Design. Washington, D.C.: National Research Council.

NRC (National Research Council). 1992a. Assessment of the U.S. Outer Continental Shelf Environmental Studies Program. II. Ecology. Washington, D.C.: National Academy Press.

NRC (National Research Council). 1992b. Assessment of the Outer Continental Shelf Environmental Studies Program. III. Social and Economic Studies. Washington, D.C.: National Academy Press.

NRC (National Research Council). 1992c. Restoration of Aquatic Ecosystems. Washington, D.C.: National Academy Press.

NRC (National Research Council). 1993a. Assessment of the U.S. Outer Continental Shelf Environmental Studies Program. IV. Lessons and Opportunities. Washington, D.C.: National Academy Press.

NRC (National Research Council). 1993b. Review of Interagency Oil Pollution Research and Technology Plan. Washington, D.C.: National Academy Press.

NSB (North Slope Borough). 1987. A Review of the Report Importance of the Eastern Alaskan Beaufort Sea to Feeding Bowhead Whales, 1985-86 (MMS 87-0037). NSB-SAC-OR-109. North Slope Borough Science Advisory Committee, Barrow, Alaska. 53 pp.

Obst, B.S., and G.L. Hunt, Jr. 1990. Marine birds feed at gray whale plumes in the Bering Sea. Auk 107:678-688.

O'Hare, M.L., L. Bacow, and D. Sanderson. 1983. Facility Siting and Public Opposition. New York: Van Nostrand Reinhold.

Øritsland, N.A., F.R. Engelhardt, F.A. Juck, R.J. Hurst, and P.D. Watts. 1981. Effect of Crude Oil on Polar Bears. Environmental Studies Report No. 24, Northern Affairs Program, Northern Environmental Protection Branch, Indian and Northern Affairs, Canada Environmental Studies, Ottawa, Ontario. 268 pp.

OSIR (Oil Spill Intelligence Report). 1993. International oil spill statistics: 1992. Oil Spill Intelligence Report Newsletter, March 18. Cutter Information Corp., Arlington, Mass.

Oliver, J.S., R.G. Kvitek, and P.N. Slattery. 1985. Walrus disturbance: Scavenging habits and recolonization of the Bering Sea benthos. J. Exp. Mar. Biol. Ecol. 91:233-246.

Overland, J.E. 1981. Marine climatology of the Bering Sea. Pp. 15-22 in The Eastern Bering Sea Shelf: Oceanography and Resources, Vol. 1, D.W. Hood and J.A. Calder, eds. National Oceanic and Atmospheric Administration, Washington, D.C.

Owens, E.H., J.R. Gould, and J. Lindstedt-Siva. 1993. Field studies to determine the ecological effects of cleanup methods on oiled shorelines. Proceedings of 1993 Oil Spill Conference. Am. Petrol. Inst. Publ. 4580:401-406.

Palinkas, L. 1990. Ethnic Differences in Coping and Depression After the *Exxon Valdez* Oil Spill. Paper presented at the 89th Annual Meeting of the American Anthropological Association, Nov., New Orleans, La. American Anthropological Association, Washington, D.C.

Paquette, R.G., and R.H. Bourke. 1974. Observations on the coastal current of arctic Alaska. J. Mar. Res. 32:7623-7630.

Paquette, R.G., and R.H. Bourke. 1981. Ocean circulation and fronts as related to ice melt-back in the Chukchi Sea. J. Geophys. Res. 86:4215-4230.

Pavia, R., W.D. Ernst, E.R. Gundlach, L.C. Thebeau, and J.L. Sadd. 1982. Sensitivity of Coastal Environments to Spilled Oil: Southern California. Office of Marine Pollution Assessment, National Oceanic and Atmospheric Administration, Boulder, Colo. 41 pp.

Payne, J.R., G.D. McNabb, Jr., L.E. Hachmeister, B.E. Kirstein, J.R. Clayton, Jr., C.R. Phillips, R.T. Redding, C.L. Clary, G.S. Smith, and G.H. Farmer. 1987. Development of a predictive model for the weathering of oil in the presence of sea ice. National Oceanic and Atmospheric Administration/Outer Continental Shelf Environmental Assessment Program (NOAA/OCSEAP) Final Reports of Principal Investigators 59(1988):147-465.

Payne, J.R., J.R. Clayton, Jr., G.D. McNabb, Jr., B.E. Kirstein, C.L. Clary, R.T. Redding, J.S. Evans, E. Reimnitz, and E.W. Kempema. 1989. Oil-ice-sediment interactions during freezeup and breakup. National Oceanic and Atmospheric Administration/Outer Continental Shelf Environmental Assessment Program (NOAA/OCSEAP) Final Reports of Principal Investigators 64(1989):1-382.

Piatt, J.F., A. Pinchuk, A. Kitayski, A.M. Springer, and S.A. Hatch. 1992. Foraging Distribution and Feeding Ecology of Seabirds at the Diomede Island, Bering Strait. U.S. Fish and Wildlife Service Final Report. OCS Study MMS 92-0041. Alaska Region, Minerals Management Service, U.S. Department of the Interior, Anchorage, Alaska.

Pritchard, R.S., and D.J. Hanzlick. 1987. Interpolation, analysis and archival of data on sea ice trajectories and ocean currents obtained from satellite-linked instruments. National Oceanic and Atmospheric Administration/Outer Continental Shelf Environmental Assessment Program (NOAA/OCSEAP) Final Reports of Principal Investigators 7-2(1990):121-238.

Ray, J.P., and F.R. Engelhardt. 1992. Produced Water: Technological/Environmental Issues and Solutions. New York: Plenum Press.

Rearic, D.M., P.W. Barnes, and E. Reimnitz. 1981. Ice Gouge Data, Beaufort Sea, Alaska, 1972-1980. USGS Open File Report 81-950. U.S. Geological Survey, Denver, Colo. 22 pp.

Reimnitz, E., and E.W. Kempema. 1984. Pack ice interaction with

Stamukhi Shoal, Beaufort Sea, Alaska. Pp. 159-183 in The Alaskan Beaufort Sea: Ecosystems and Environment, P.W. Barnes, D.M. Schell, and E. Reimnitz, eds. New York: Academic Press.

Richardson, W.J., and C.I. Malme. 1993. Man-made noise and behavioral responses. Pp. 631-700 in The Bowhead Whale, Society for Marine Mammalogy Special Publ. No. 2, J.J. Burns, J.J. Montague, and C.J. Cowles, eds. Lawrence, Kan.: Allen Press.

Roseneau, D.G., and D.R. Herter. 1984. Marine and costal birds. Pp. 81-115 in Proceedings of a Synthesis Meeting: The Barrow Arch Environment and Possible Consequences of Planned Offshore Oil and Gas Development, J.C. Truett, ed. National Oceanic and Atmospheric Administration/Outer Continental Shelf Environmental Assessment Program (NOAA/OCSEAP), Anchorage, Alaska.

Sanders, H.L., J.F. Grassle, G.R. Hampson, L.S. Morse, S. Garner-Price, and C.C. Jones. 1980. Anatomy of an oil spill: Long-term effects from the grounding of the barge Florida off West Falmouth, Massachusetts. J. Mar. Res. 38:265-380.

Sanderson, T.J., ed. 1987. Working Group on Ice Forces: Third State of the Art Report. CRREL Special Rep. 87-17. Cold Regions Research and Engineering Laboratory, U.S. Army Corps of Engineers, Hanover, N.H.

Sanderson, T.J. 1988. Ice Mechanics and Risks to Offshore Structures. Boston: Graham and Trotman.

Schauer, A.E.S. 1993. Association between seabirds and water masses in the northern Bering Sea. Pp. 388-396 in Results of the Third Joint US-USSR Bering and Chukchi Seas Expedition (BERPAC), Summer 1988, P.A. Nagel, ed. U.S. Fish and Wildlife Service, U.S. Department of the Interior, Washington, D.C.

Schwartz, S., P. White, and R. Hughes. 1985. Environmental threats, communities, and hysteria. J. Public Health Policy 6:58-75.

Scudder, T. 1973. The human ecology of big projects: River basin development and resettlement. Pp. 45-55 in Annual Review of Anthropology, Vol. 2. Palo Alto, Calif.: Annual Reviews.

Scudder, T. 1982. No Place to Go: Effects of Compulsory Relocation on Navajos. Institute for the Study of Human Issues, Philadelphia, Pa.

Scudder, T., and E. Colson. 1972. The Kariba Dam project: Resettlement and local initiative. In Technology and Social Change, H.R. Bernard and P. Pelto, eds. New York: Macmillan.

Seaman, G.A., and J.J. Burns. 1981. Preliminary results of recent studies of belukhas in Alaskan waters. Rep. Int. Whal. Comm. 31:567-574.

Seaman, G.A., L.F. Lowry, and K.J. Frost. 1982. Foods of belukha whales (*Delphinapterus leucas*) in western Alaska. Cetology 44:1-19.

Sellmann, P.V., and D.M. Hopkins. 1984. Subsea permafrost distribution on the Alaskan shelf. Pp. 75-82 in Final Proceedings of the Fourth International Permafrost Conference, July 17-22, 1983, Fairbanks, Alaska. Washington, D.C.: National Academy Press.

Sergy, G.A. 1985. The Baffin Island Oil Spill (BIOS) project: A summary. Proceedings of 1985 Oil Spill Conference. Am. Petrol. Inst. Publ. 4385:571-575.

Shapiro, L.H., and J.J. Burns. 1975. Satellite observations of sea ice movement in the Bering Strait region. Pp. 379-386 in Climate of the Arctic, G. Weller and S.A Bowling, eds. Twenty-fourth Alaska Science Conference, Aug. 15-17, 1973. Geophysical Institute, University of Alaska, Fairbanks.

Shapiro, L.H., and P.W. Barnes. 1991. Correlation of nearshore ice movement with seabed gouges near Barrow, Alaska. J. Geophys. Res. 96:16979-16989.

Shaughnessy, P.D., and F.H. Fay. 1977. A review of the taxonomy and nomenclature of North Pacific harbour seals. J. Zool. (London) 182:385-419.

Slovic, P., J. Flynn, and M. Layman. 1991. Perceived risk, trust, and the politics of nuclear waste. Science 254:1603-1607.

Sowls, A.L., S.A. Hatch, and C.J. Lensink. 1978. Catalog of Alaskan Seabird Colonies. U.S. Fish and Wildlife Service FWS/OBS-78/78. Biology Services Program, U.S. Fish and Wildlife Service, Anchorage, Alaska.

Spies, R.B. 1983. Natural submarine petroleum seeps. Oceanus 26(3):24-29.

Springer, A.M., and D.G. Roseneau. 1985. Copepod-based food webs: Auklets and oceanography in the Bering Sea. Mar. Ecol. Prog. Ser. 21:229-237.

Springer, A.M., D.G. Roseneau, and M. Johnson. 1979. Ecological studies of colonial seabirds at Cape Thompson and Cape Lisburne, Alaska. Environmental Assessment of the Alaskan Continental Shelf, Annual Reports 2:517-574.

Springer, A.M., E.C. Murphy, D.G. Roseneau, and M.I. Springer. 1982.

Population status, reproductive ecology, and trophic relationships of seabirds in northwestern Alaska. Final report to the National Oceanic and Atmospheric Administration/Outer Continental Shelf Environmental Assessment Program (NOAA/OCSEAP), Juneau, Alaska. 129 pp.

Springer, A.M., D.G. Roseneau, E.C. Murphy, and M.I. Springer. 1985. Population and trophic studies of seabirds in the northern Bering and eastern Chukchi seas, 1982. National Oceanic and Atmospheric Administration/Outer Continental Shelf Environmental Assessment Program (NOAA/OCSEAP) Final Reports of Principal Investigators 30(1985):59-126. Available as NTIS PB 86-134756/AS.

Springer, A.M., E.C. Murphy, D.G. Roseneau, C.P. McRoy, and B.A. Cooper. 1987. The paradox of pelagic food webs in the northern Bering Sea. I. Seabird food habits. Continental Shelf Res. 7:895-911.

Springer, A.M., D.G. Roseneau, E.C. Murphy, and M.I. Springer. 1989. Environmental controls of marine food webs: Food habits of seabirds in the eastern Chukchi Sea. Can. J. Fish Aquat. Sci. 41:1202-1215.

SR (Special Report) 6. Lawrence Johnson and Associates, Inc. 1985. Alaska OCS Studies Program: Review of Outer Continental Shelf Economic and Demographic Impact Modeling for Rural Alaska: Proceedings of a Workshop. OCS Study MMS 85-0080. Alaska Region, Minerals Management Service, U.S. Department of the Interior, Anchorage, Alaska. Available as NTIS PB 87-204699/AS.

Stirling, I. 1983. Conservation problems with Arctic Ocean mammals. N.Z. Antarctic Rec. 5(1):44-45.

Stirling, I., and W. Calvert. 1983. Environmental threats to marine mammals in the Canadian Arctic. Polar Rec. 21:433-449.

Stoker, S., and I.I. Krupnik. 1993. Subsistence whaling. Pp. 579-629 in The Bowhead Whale, Society for Marine Mammalogy Special Publ. No. 2, J.J. Burns, J.J. Montague, and C.J. Cowles, eds. Lawrence, Kans.: Allen Press.

Stringer, W.J., J. Zender-Romick, and J.E. Groves. 1982. Width and Persistence of the Chukchi Polynya. Report to National Oceanic and Atmospheric Administration/Outer Continental Shelf Environmental Assessment Program (NOAA/OCSEAP), Anchorage, Alaska. 22 pp.

Sveum, P., and C. Bech. 1991. Burning of oil in snow: Experiments and implementation in a Norsk hydro drilling contingency plan. Pp. 399-419 in Proceedings of the Fourteenth Arctic and Marine Oilspill Program Technical (AMOP) Seminar, Vancouver, British Columbia, June 12-14.

Environment Canada, Ottawa, Ontario.

Tetra Tech. 1985. Oil Spill Cleanup: Options for Minimizing Adverse Ecological Impacts. Publ. No. 4435. American Petroleum Institute, Washington, D.C. 600 pp.

Thomas, D.R. 1983. Potential oiled ice trajectories in the Beaufort Sea. National Oceanic and Atmospheric Administration/Outer Continental Shelf Environmental Assessment Program (NOAA/OCSEAP) Final Reports of Principal Investigators 40(1986):295-402.

Thomas, C.P., T.C. Doughty, D.D. Faulder, W.E. Harrison, J.S. Irving, H.C. Jamison, and G.J. White. 1991. Alaska Oil and Gas: Energy Wealth or Vanishing Opportunity? DOE/ID/01570-H1. Office of Fossil Energy, U.S. Department of Energy, Washington, D.C. Available as NTIS DE 91006616.

Thorndike, A.S. 1986. Kinematics of sea ice. Pp. 489-549 in The Geophysics of Sea Ice, N. Untersteiner, ed. NATO ASI Series B, Physics, Vol. 146. New York: Plenum Press.

Thorson, G. 1950. Reproductive and larval ecology of marine bottom invertebrates. Biol. Rev. 25:1-45.

Tickamyer, A.R., and C.H. Tickamyer. 1988. Gender and poverty in central Appalachia. Soc. Sci. Q. 69(4):874-91.

Timco, G.W., ed. 1989. Working Group on Ice Forces: Fourth State of the Art Report. CRREL Special Rep. 89-5. Cold Regions Research and Engineering Laboratory, U.S. Army Corps of Engineers, Hanover, N.H.

TR (Technical Report) 85. Kruse, J.A., M. Baring-Gould, W. Schneider, J. Gross, G. Knapp, and G. Sherrod. 1983. A Description of the Socioeconomics of the North Slope Borough. A final report by the Institute of Social and Economic Research, University of Alaska for the Alaska OCS Region, Minerals Management Service, U.S. Department of the Interior, Anchorage, Alaska. Available as NTIS PB 87-189338.

TR (Technical Report) 100. Knapp, G., and W. Nebesky. 1983. Economic and Demographic Systems Analysis, North Slope Borough. A final report by the Institute of Social and Economic Research, University of Alaska for the Alaska OCS Region, Minerals Management Service, U.S. Department of the Interior, Anchorage, Alaska. Available as NTIS PB 87-207086.

TR (Technical Report) 117. Smythe, C.W., R. Worl, S. Langdon, T. Lonner, and T. Brelsford. 1985. Monitoring Methodology and Analysis of North Slope Institutional Response and Change, 1979-1983. OCS

Study MMS 85-0072. A final report by the Chilkat Institute for the Alaska OCS Region, Minerals Management Service, U.S. Department of the Interior, Anchorage, Alaska. Available as NTIS PB 87-204715.

TR (Technical Report) 120. Knapp, G., S. Colt, and T. Henley. 1986. Economic and Demographic Systems of the North Slope Borough: Beaufort Sea Lease Sale 97 and Chukchi Sea Lease Sale 109, Vol. 1, Description and Projections; Vol. 2, Data Appendices. OCS Study MMS 86-0019. A final report by the Institute of Social and Economic Research, University of Alaska for the Alaska OCS Region, Minerals Management Service, U.S. Department of the Interior, Anchorage, Alaska. Available as NTIS PB 87-205241.

TR (Technical Report) 125. Worl, R., and C.W. Smythe. 1986. Barrow: A Decade of Modernization. (The Barrow Case Study). OCS Study MMS 86-0088. A final report by the Chilkat Institute for the Alaska OCS Region, Minerals Management Service, U.S. Department of the Interior, Anchorage, Alaska. Available as NTIS PB 87-204673.

TR (Technical Report) 133. Stephen R. Braund & Associates with the Institute of Social and Economic Research, University of Alaska. 1988. North Slope Subsistence Study: Barrow, 1987. OCS Study MMS 88-0080. A final report to the Alaska OCS Region, Minerals Management Service, U.S. Department of the Interior, Anchorage, Alaska. Available as NTIS PB 91-105569.

USGS/MMS (U.S. Geological Survey/Minerals Management Service). 1989. Estimates of Undiscovered Conventional Oil and Gas Resources in the United States—A Part of the Nation's Energy Endowment. U.S. Geological Survey, Denver, Colo.

USGS (U.S. Geological Survey). 1992a. 1991 Annual Report on Alaska's Mineral Resources. USGS Circular 1072. U.S. Geological Survey, Denver, Colo.

USGS (U.S.Geological Survey). 1992b. Map I-1182-H. U.S. Geological Survey, Denver, Colo.

Vaudrey, K.D. 1977. Determination of mechanical sea ice properties by large scale-field beam experiments. Pp. 529-543 in POAC 77: Proceedings of the International Conference on Port and Ocean Engineering Under Arctic Conditions, Sept. 26-30, St. John's, Newfoundland. St. Johns, Newfoundland: Memorial University.

Vivitrat, V., and T.R. Kreider. 1981. Ice force prediction using a limited driving force approach. Pp. 471-485 in 1981 Proceedings of the 13th Annual Offshore Technology Conference, Vol. 3. Offshore Technology

Conference, Houston, Tex.

Walsh, J.E., and C.M. Johnson. 1979. An analysis of arctic sea ice fluctuations, 1953-1977. J. Phys. Oceanogr. 9:580-591.

Wang, A.T. 1990. Numerical simulations for rare ice gouge depths. Cold Regions Sci. Technol. 19(1):19-32.

Watson, G.E., and G.J. Divoky. 1972. Pelagic Bird and Mammal Observations in the Eastern Chukchi Sea, Early Fall 1970. U.S. Coast Guard Oceanogr. Rep. CG-50. U.S. Coast Guard, Washington, D.C.

Weber, B.A., and R.E. Howell, eds. 1982. Coping with Rapid Growth in Rural Communities. Boulder, Colo.: Westview Press.

Weeks, W.F., and R.L. Handy. 1970. Icebreaking Capability, Test Voyage Report: 1970 S.S. Manhattan Full Scale Ice Breaking Tests. Humble [Exxon] Oil and Refining Co.

Weeks, W.F., and G. Weller. 1984. Offshore oil in the Alaskan Arctic. Science 225:371-378.

Weeks, W.F., P.W. Barnes, D.M. Rearic, and E. Reimnitz. 1984. Some probabilistic aspects of ice gouging on the Alaskan shelf of the Beaufort Sea. Pp. 213-236 in The Alaskan Beaufort Sea: Ecosystems and Environment, P.W. Barnes, D.M. Schell and E. Reimnitz, eds. New York: Academic Press.

Weeks, W.F., W.B. Tucker III, and A.W. Niedoroda. 1985. A numerical simulation of ice gouge formation on the shelf of the Beaufort Sea. Pp. 393-407 in POAC 85: Proceedings of the International Conference on Port and Ocean Engineering Under Arctic Conditions, Narssarssuaq, Greenland. Horsholm, Denmark: Danish Hydraulic Institute.

Westree, B. 1977. Biological criteria for the selection of cleanup techniques in salt marches. Proceedings of 1977 Oil Spill Conference. Am. Petrol. Inst. Publ. 4284:231-235.

Wiseman, W.J., Jr., and L.J. Rouse, Jr. 1980. A coastal jet in the Chukchi Sea. Arctic 33:21-29.

Woodby, D.A., and G.J. Divoky. 1982. Spring migration of eiders and other waterbirds at Point Barrow, Alaska. Arctic 35:403-410.

Wyman, M. 1950. Deflections of an infinite plate. Can. J. Res. Ser. A 28:293-302.

Zeh, J.E., C.W. Clark, J.C. George, D. Withrow, G.M. Carroll, and W.R. Koski. 1993. Current population size and dynamics. Pp. 409-489 in The Bowhead Whale, Society for Marine Mammalogy Special Publ. No. 2, J.J. Burns, J.J. Montague, and C.J. Cowles, eds. Lawrence, Kans.: Allen Press.

APPENDIX A

MMS'S OCS OIL AND GAS DECISION PROCESS

(Prepared by NRC and MMS staff)

ANALYSES REQUIRED BY SECTION 18 OF THE OCS LANDS ACT

The leasing program consists of a schedule of proposed lease sales that indicates as precisely as possible the timing, size, and location of leasing activity that the secretary of the interior determines will meet the nation's energy needs for the 5-year period following the schedule's approval (MMS, 1991a [OCS National Compendium through October 1990]). The current 5-year plan, approved in July 1992, is for mid-1992 through mid-1997. Section 18 of the Outer Continental Shelf Lands Act (OCSLA, § 43 U.S.C. 1344) requires that MMS analyze the different OCS regions to consider the following before it determines where the leased sites will be (MMS, 1992c [1992-1997 Proposed Final Comprehensive Program]):

- Geographical, geological, and ecological characteristics.
- Equitable sharing of benefits and environmental risks of development.
- Location with respect to regional and national energy needs and energy markets.
- Location with respect to other users of the sea.
- Petroleum industry interest.

- Laws, goals, and policies of the affected states.
- Relative environmental sensitivity and marine productivity.
- Relevant environmental and predictive information.

The U.S. portion of the OCS is divided into 26 planning areas (a geographic subdivision of the OCS that is used for planning and administrative purposes) to facilitate the Proposed Final Comprehensive Program's analysis and program management as required by Section 18. The Section 18 analysis is to determine the value to the nation as a whole of all the estimated oil and gas resources in each of 14 OCS planning areas proposed for leasing consideration in MMS's July 1992 Proposed Final Comprehensive Program. Cost-benefit analysis is used to determine a lease site's value to the nation. The analysis compares the net economic value of oil and gas to be produced against environmental externalities that can be quantified as social costs (costs not borne directly by oil and gas producers). The OCS program areas (an area within one or more planning areas that is proposed for leasing consideration in the Proposed Final Comprehensive Program (1992-1997); there are 14 program areas made up of portions of 15 planning areas) are then ranked in groups in order of descending value (MMS, 1992c). Group 1 has the highest net social value; Group 5 has the lowest net social value. The Beaufort Sea and Chukchi Sea planning areas are in Group 2, and the Navarin Basin is in Group 3.

In addition to net social value, MMS's Proposed Final Comprehensive Program analyzes five other components. The first is the additional costs and benefits, such as indirect benefits related to price effects and national income effects, the potential costs of proposed environmental regulations, and the nonquantifiable costs analyzed in final Environmental Impact Statements (EISs); the findings of this analysis were not grouped in the current 5-year plan.

Second, an analysis of the relative marine productivity and environmental sensitivity is broken into three components: coastal habitats, marine habitats, and marine biota. Estimated total primary productivity and average primary productivity are given for each planning area, rather than for each program area, because of difficulty in applying the information to smaller areas. The current 5-year plan estimates that the Navarin Basin has moderate primary productivity and that the Chukchi and Beaufort seas have low primary productivity relative to other areas.

The third analysis is of the equitable sharing of development benefits and environmental risks among OCS regions. This analysis presents the means of distributing developmental benefits and compensating costs associated with the OCS oil and gas program. It also includes a review of the effects of constraints beyond the control of the Department of the Interior, such as congressional moratoria, on the program.

The fourth analysis reviews industry interest in leasing, exploration, and development of oil and gas resources in the program areas.

The fifth analysis comes directly from Section 18 language. The secretary is to choose the leasing program that "he determines will best meet the national energy needs for the 5-year period following its approval or reapproval" and "to the maximum extent practicable, to obtain a proper balance between the potential for environmental damage, the potential for discovery of oil and gas, and the potential for adverse impact on the coastal zone." In this analysis, the secretary is to evaluate and balance four considerations. The first is developmental benefits, such as increased national income, economic activity, and job opportunities and reduced national deficits in international trade and in the federal budget. The second is the potential for environmental damage from such activities as placement of infrastructure, operations, and accidents. The third consideration is more general and includes aesthetics and special concern for marine mammals or endangered species. The fourth consideration is the comments of interested or affected parties.

AREA EVALUATION AND DECISION PROCESS

In the first step in the oil and gas leasing process (called the Area Evaluation and Decision Process), MMS prepares a proposed 5-year oil and gas leasing schedule for review by the public, industry, states, and other affected agencies, as well as by other agencies within the department, such as the National Park Service (NPS) and the Fish and Wildlife Service (FWS). The Information Base Review process gathers specific information on geology, resource estimates, the environment, and economics. The specific information on geology and resource reports is requested from the states, other federal agencies, NPS, and FWS, and is used to define the area in which oil or gas could be found. Once the potential areas of resources have been defined, a public notice of Request for Interest and Comments is published in the *Federal Register*. The degree of response to a request indicates the degree of industry and public interest in an area.

Next, a Call for Information and Nominations and a Notice of Intent to

Prepare an EIS are published in the *Federal Register*. The call solicits comments from the oil industry, government agencies, environmental groups, and the general public. The notice of intent announces the scoping (information gathering and evaluation) process that will be followed for the EIS. During this process, important issues that deserve study in an EIS are identified. A 45-day comment period follows. After the call closes, the comments from the public, industry, states, and other federal agencies, including NPS and FWS, are reviewed. In addition, MMS evaluates potential impacts on multiple uses—fishing, tourism, shipping, etc.—in the EIS process. An area for proposed leasing is identified at this point and submitted to the secretary for approval.

If the area is approved by the secretary, the scoping process ensues. It is open to input from the public, industry, states, and other federal agencies, including NPS and FWS. The process allows for early identification of important issues that deserve review in an EIS. Afterwards, a draft EIS and a Proposed Notice of Sale are prepared. A 60-day comment period and public hearings on the draft EIS and public notice are publicized by a Notice of Availability in the *Federal Register*.

A final EIS and a Proposed Action and Alternatives Memorandum, which are submitted for the secretary's approval, are prepared in the appropriate regional office. The memorandum's purpose is to assist the secretary in determining whether to proceed with, delay, or cancel further development and analysis of a leasing proposal. If the secretary decides to proceed, a tentative sale decision is made; the proposed sale is publicized by a Notice of Availability in the *Federal Register*. A 60-day comment period and public hearings follow. The state governors are notified of the proposed sale and are asked for input. During the 60-day comment period, the secretary holds a meeting to decide on areas, terms, and conditions.

If the secretary gives final approval for the sale, a final Notice of Sale is published in the *Federal Register* at least 30 days before the sale. Industry bids are received within schedules identified in the notice. On the day of the sale, the bids are opened and read publicly. No decision on the adequacy of the bids is made at that time. The Department of Justice and the Federal Trade Commission then review the bids. The sale results also are reviewed in the regional office, which then determines whether to accept a bid. If a bid is accepted, a lease is issued by the regional office to the highest bidder (MMS, 1991a [OCS National Compendium through October 1990]).

After a lease has been issued, all exploration, development, and production activities, except for preliminary activities, must be conducted in accordance with an Exploration Plan (EP) or a Development and Production Plan (DPP) approved by MMS. Preliminary activities, including geophysical, geological, and other surveys, can be conducted to gather information for preparation of an EP or DPP. These surveys may not result in physical penetration of the seabed more than 500 feet nor may they result in any significant harm to natural resources of the OCS.

EXPLORATION PLAN

Once a lease has been issued, the lessee may submit an EP, which is reviewed and commented on by the state governor, with input from federal agencies. The EP provides a concise description of the proposed offshore operations and must be modified by the lessee if significant changes are made to the lessee's proposal. The EP is based on information collected by the lessee during preliminary activities, including, but not limited to, geophysical, geological, cultural, and biological surveys. It is reviewed by MMS to ensure that it is complete within 10 working days of receipt. If it is not complete, it is returned to the lessee for modification or supplemental information. Once the EP is complete, it is deemed submitted. Within 2 working days, the plan is sent to the affected states' governors, the Coastal Zone Management agencies, NPS and FWS, and local governments that request copies.

While the EP is being reviewed by state and other agencies, MMS conducts a technical review of the proposal. Concurrently, MMS regional staff analyze all activities proposed in the EP as required by the National Environmental Policy Act in an Environmental Assessment (EA). The EA reviews whether the proposed EP will have a significant impact on the environment. If no significant impact is identified, the plan proceeds through the approval process.

If the EP is rejected, it goes back to the lessee for modification or for supplemental information, and the process starts again. The plan may be resubmitted if conditions change. Leases can be canceled at this point in accordance with OCSLA § 43 U.S.C. 1334, although no such cancelation has occurred to date.

DEVELOPMENT AND PRODUCTION PLAN

After the EP is approved, the lessee must submit a DPP, which describes the specific work to be performed; the drilling vessels, platforms, pipelines, and facilities to be used; the location and depth of proposed wells; geological and physical data; environmental safeguards and standards to be met; and a schedule of development and production activities. If the DPP is deemed complete by MMS, it must be reviewed by the affected state and federal agencies, NPS and FWS, and the public, which is notified through the *Federal Register*. The technical review is similar to that for an EP, except that the review must be completed within 120 days after the plan is deemed completed. If it is determined that an EIS must or should be prepared, the 120-day restriction does not apply.

ENVIRONMENTAL IMPACT STATEMENT

An EIS addresses the environmental effects of the proposal, including seismic risks, descriptions of areas of high ecological sensitivity and of hazardous bottom conditions, and the uses of new or unusual technology. It is prepared either when MMS determines that approval of a DPP constitutes a major federal action that would significantly affect the quality of the environment or for an initial DPP activity in a frontier area (i.e., an area in which no oil or gas development or production has occurred). When an EIS is deemed necessary, particular consideration is given to addressing the significant adverse effects on the marine, coastal, or human environment resulting from the construction of new onshore and offshore facilities. Cumulative impacts are also considered.

Once the EA has been reviewed, the DPP is approved if the Coastal Zone Management agencies have no objections. Coast Guard and U.S. Army Corps of Engineers permits are sought at this point. If approved and the permits are issued, a platform verification review is done in some cases; for example, in newly permitted areas where there are hazardous conditions, such as ice, and other climatic hazardous or oceanic conditions.

APPLICATION FOR PERMIT TO DRILL

Next, the Application for Permit to Drill (APD) is submitted by the lessee. If it is issued, a National Pollution Discharge Elimination System (NPDES) permit is sought from the Environmental Protection Agency.

MMS then determines whether the APD is technically acceptable. An APD must be filed and approved each time a well is proposed to be drilled, deepened, or plugged and abandoned. Extensive details about the drilling program are required, including information about the blowout prevention system and casing, cementing, and drilling-mud programs. MMS uses this information to evaluate the operational safety and pollution prevention measures of a proposed operation. Conditions of approval are often attached to the APD as a result of MMS's review of it.

If the APD is approved, the lessee is given an Exploration Drilling Permit to drill a test well. The lessee must also obtain a Coast Guard permit, a U.S. Army Corps of Engineers Navigation Permit, and an EPA NPDES permit. Once the required permitting is complete, exploratory drilling may begin.

Next, the lessee must submit a Pipeline Permit Application. MMS prepares an EA, including off-lease right-of-way, which must take Department of Transportation regulations into consideration and undergo review by NPS and FWS. After the EA has been prepared, the Coastal Zone Management agencies notify MMS of any objections they have. If there are no objections, MMS approves a pipeline permit. Next the lessee must obtain a U.S. Army Corps of Engineers construction permit, after which pipeline construction and production can begin.

Once production has begun, the lessee must report its royalties and send monthly production reports to MMS's Office of Official Records, Title Actions. After production has ceased, relinquishment occurs. Shutdown requires well abandonment, platform removal, and clearance of the seafloor within one year.

TERMINATION OF OPERATIONS

The lessee must obtain MMS approval before beginning to permanently or temporarily abandon operations. The lessee must adhere to strict standards for removing platforms and subsea wellheads, and it must clear the seafloor of all other obstructions. MMS must consult with other federal agencies to ensure that abandonment operations do not harm natural resources in the vicinity of these operations.

Abbreviations and Acronyms

AEDP	Area Evaluation and Decision Process
ANWR	Arctic National Wildlife Refuge
APD	Application for Permit to Drill
ARCO	Atlantic Richfield Company
ASTM	American Society for Testing and Materials
BEOA	MMS's Branch of Environmental Operations and Analysis
BIOS	Baffin Island Oil Spill Project
BLM	Bureau of Land Management
CDP	Common depth point
COE	U.S. Army Corps of Engineers
COST	Continental Offshore Stratigraphic Test
DNAG	Decade of North American Geology
DPP	Development and Production Plan
EA	Environmental Assessment
EIS	Environmental Impact Statement

EP	Exploration Plan
EPA	U.S. Environmental Protection Agency
ESA	Endangered Species Act
ESP	Environmental Studies Program
FWS	U.S. Fish and Wildlife Service
GCM	General circulation model
GESAMP	International Maritime Organization Group of Experts on the Scientific Aspects of Marine Pollution
GPS	Global Positioning System
MMPA	Marine Mammal Protection Act
MMS	Minerals Management Service
NEPA	National Environmental Policy Act
NMFS	National Marine Fisheries Service
NOAA	National Oceanic and Atmospheric Administration
NPDES	National Pollution Discharge Elimination System
NPRA	National Petroleum Reserve, Alaska
NPS	National Park Service
NRC	National Research Council
OCS	Outer continental shelf
OCSEAP	Outer Continental Shelf Environmental Assessment Program
OCSLA	Outer Continental Shelf Lands Act
OMB	Office of Management and Budget
ONR	Office of Naval Research
OSRA	Oil spill risk analysis

PAAM	Proposed Action and Alternatives Memorandum
PRESTO	Probalistic Resource Estimates—Offshore
RTWG	Regional Technical Working Group
SAR	Synthetic aperture radar
SESP	Socioeconomics Studies Program
SID	Secretarial issue document
Sv	Sverdup (1 Sv = 10^6 m^3/s)
TAPS	Trans-Alaska Pipeline System
USCG	U.S. Coast Guard
USGS	U.S. Geological Survey

GLOSSARY

Advection   Usually horizontal transport of air or water, as a current.

Anticline   A fold or arch of stratified rock whose layers bend downward from the crest and whose core contains stratigraphically older rocks.

Baroclinicity   A condition in which surfaces of equal pressure are inclined to surfaces of equal density.

Benthos   The organisms that live in or on the bottom of water bodies. Hence benthic.

Brookian   A local term indicating geological age applied to northern Alaska.

Cambrian   The earliest period of the Paleozoic era, thought to have covered the span of time between 570 and 500 million years ago.

Clastic   Made up of fragments of pre-existing rocks.

Coriolis force   Apparent force due to Earth's rotation that deflects moving objects to the right in the northern hemisphere and to the left in the southern hemisphere. The magnitude of force increases as one moves toward the poles.

Cretaceous

The final period of the Mesozoic era (after the Jurassic and before the Tertiary period of the Cenozoic era), thought to have covered the time span between 135 and 65 million years ago.

Cryology

The study of snow and ice.

Danian

The lowest (oldest) division of the Tertiary period.

Deformation

A general term for the processes of folding, faulting, shearing, compression, or extension of a material such as a rock or sea ice.

Devonian

A period of the Paleozoic era (after the Silurian and before the Mississippian), thought to have covered the span of time between 400 and 345 million years ago.

Eocene

An epoch of the early Tertiary period, after the Paleocene and before the Oligocene.

Foredeep

A deep depression in the ocean bottom fronting a mountainous land area.

Frazil ice

Ice crystals or granules sometimes resembling slush that are formed in supercooled turbulent water. When these ice crystals float to the surface, they form a layer of fine-grained, granular ice.

Geostrophic

Resulting from a balance between the pressure gradient force and the Coriolis force (as in the geostrophic wind).

Graben

A depressed segment of the Earth's crust bounded on at least two sides by faults and generally of greater length than width.

Halocline

A location in the water column where there is a usually sharp vertical gradient in salinity.

Hydrography	The science of describing the waters of the Earth's surface comprising the study and mapping of their forms and physical features; i.e., the contours of the sea bottom, shallows, winds, tides, and currents.
Lead	A long narrow open water opening passing through a region of sea ice navigable by surface vessels and air-breathing mammals. It is often said that the difference between a crack and a lead is that one can row a boat down a lead.
Littoral	Of, relating to, on, or near a shore, especially of the sea.
Maestrichtian	Of, pertaining to, or designating the uppermost division of the Upper Cretaceous in Europe. It immediately underlies the Danian stage.
Marine boundary layer	The region of the atmosphere over the sea where heat, moisture, and momentum are exchanged by turbulent transfer between the sea surface and the air. The layer's depth varies, typically from a few hundred to a thousand meters.
Miocene	An epoch of the Tertiary period, after the Oligocene and before the Pliocene.
Oligocene	An epoch of the Tertiary period after the Eocene and before the Miocene.
Orographic	Of or relating to mountains, especially with respect to their location, distribution, and accompanying phenomena (e.g., orographic lifting of air).
Pelagic	Of, relating to, or living in the open sea or in midwater; not associated with the benthos.
Pleistocene	An epoch of the Quaternary period, after the Pliocene of the Tertiary and before the Holocene.

Pliocene
: An epoch of the Tertiary period, after the Miocene and before the Pleistocene

Polynya
: A large area of open water or atypically thin ice in a location that is surrounded by sea ice and where climatologically sea ice should occur. The term is commonly applied to a persistent feature.

Reservoir rock
: A subsurface volume of porous and permeable rock in which oil or gas has accumulated.

Rift
: A long, narrow continental trough bounded by normal faults.

Silurian
: A period of the Paleozoic era, thought to have covered the span of time between 440 and 400 million years ago.

Slump
: The sliding down of a mass of sediment shortly after its deposition on an underwater slope, especially the downslope flow of soft unconsolidated marine sediments at the head or along the side of a submarine canyon.

Source rock
: The geological formation in which oil and gas originate.

Stratigraphic sequence
: A chronological succession of sedimentary rocks from older below to younger above, essentially without interruption.

Subsea completion
: Completion of an oil well with the wellhead at the sea floor.

Synthetic Aperture Radar
: A technique in which numerical procedures are used to generate high-resolution radar images. For the procedure to work, the radar move at a known rate (e.g., on an airplane or satellite) over the terrain to be imaged.

Taxon (pl. taxa)
: A group of organisms classified according to presumed evolutionary relationship. Examples of taxa include phylum, family, genus, species.

Tectonic   Of or relating to deformation of the Earth's crust, the forces involved in or producing such deformation, and the resulting rock structures and external forms.

Tertiary   The first period of the Cenozoic era, thought to have covered the span of time between 65 and 2 million years ago.

Transgression   The spread of the sea over land areas; any change that brings offshore, deep-water environments to areas occupied by nearshore, shallow-water conditions, or that shifts the boundary between marine and nonmarine deposition outward from the center of a marine basin.

Biographical Information on Committee Members

Charles G. Groat (chair) is executive director of the Center for Coastal, Energy, and Environmental Resources at Louisiana State University (1992-). Formerly, he was executive director of the American Geological Institute (1990-1992), was Assistant to the Secretary of the Louisiana Department of Natural Resources (1978-1990), and was State Geologist and Director of the Lousiana Geological Survey. He is a member of the NRC Committee on Earth Resources (1991-), and was a member of the NRC committee on Undiscovered Oil and Gas Resources (1988-1990). He represents Louisiana on the Outer Continental Shelf (OCS) Policy Committee that advises the Secretary of the Interior on OCS oil and gas leasing policy and environmental matters, and was chair (1990-1992). He is a member of AAAS, the American Association of Petroleum Geologists, the Association of American State Geologists (President, 1987-1988), and the Geological Society of America. His research involves coastal and energy resources research, including direction of federally funded programs in coastal environmental mapping, wetland mitigation, and oil and gas resources. He received a BA in 1962 from the University of Rochester, an MS in 1967 from the University of Massachusetts, and a PhD (geology) in 1970 from the University of Texas-Austin.

John J. Amoruso is an independent geologist and general partner with Amoruso Petroleum Company (1969-). He explores for oil and gas in the onshore petroleum basins of the United States. He is a member of the American Association of Petroleum Geologists (President, 1983-1984), the Society of Independent Earth Scientists (President, 1980-1981), and is a fellow of the Geological Society of America. He is President of the American Geological Institute for 1993-1994. He was a member of the NRC Committee on Undiscovered Oil and Gas Resources (1988-1990). He received a BS in 1952 from Tufts College and an MS (geology) in 1957 from University of Michigan.

John C. Crowell is emeritus professor of geology at the University of California, Santa Barbara (1987-). He is a member of the National Academy of Sciences. He is a fellow of the Academy of Arts and Sciences, and is a member of the Geological Society of America, the American Association of Petroleum Geologists, the American Geophysical Union, and Sigma Xi. His research includes structural and general geology, tectonics, and paleoclimatology of ancient ice ages. He received a BS in 1939 from the University of Texas, an MA (oceanographic meteorology) in 1946 and PhD (geology) in 1947 from the University of California, Los Angeles.

Rainer Englehardt is vice president for research & development at Marine Spill Response Corporation. He was director of the Pollution Control Division of the Environmental Protection Branch of Canada's Oil and Gas Lands Administration where he was responsible for the development of marine environmental policies regulating oil and gas activities in Canada, environmental licensing of exploration and development activities, and research and development to support criteria for environmental protection. He received a BS in 1965 from University of Western Ontario, a MS in 1967 from the University of British Columbia, and a PhD (environmental physiology) in 1972 from the University of Guelph.

William Freudenburg is a professor in the Department of Rural Sociology at the University of Wisconsin (1991-, associate professor 1986-1991). He was a panelist for the NRC panel on the future of nuclear power (1984), and a participant in the NRC workshop to review the Emergency Planning and Community Right to Know Act of 1986 (1990). He was a member of MMS's Scientific Advisory Panel on the Outer Continental Shelf (1983-1990) and was chair of its socioeconomic subcommittee (1986-1990). He is a secretary of AAAS (1986-), and is a member of the American Sociological Association, the Society for Applied Anthropology, the International Association for Impact Assessment, and is currently Vice President of the Rural Sociological Society. His research interests include social impact assessment, measurement of social science variables, estimation of risk, and resource development. He received a BA in 1974 from the University of Nebraska-Lincoln, an MA in 1976, MPhil in 1977, and PhD (sociology) in 1979 from Yale University.

Kathryn J. Frost is a marine mammals biologist with the Alaska Department of Fish and Game (1975-). Her studies include assessing the environmental impact of offshore oil and gas exploration and development in northern Alaska through research on various mammal species (beluga and bowhead whales; ringed, bearded and spotted seals; walrus). She is a fellow of the Arctic Institute of North America and a charter member of the Society for Marine Mammalogy. She received her BS in 1970 from Tulane University, and an MS (marine sciences) in 1977 from University of California-Santa Cruz.

Christopher J. R. Garrett is Lansdowne Professor at the University of Victoria (1991-). Previously, he was a professor in the department of oceanography at Dalhousie University (1977-1991). He was a member of the NRC Earth Sciences Grant Committee (1976). His research interests are fundamental ocean processes, particularly those related to mixing, including boundary mixing and other fluid dynamical processes occurring at the sloping sides of ocean basins or continental shelves. He has also worked with the development of resources off the east coast of Canada. He received a BA in 1965 and a PhD (geophysical fluid dynamics) in 1968 from Cambridge.

George L. Hunt, Jr. is a professor of behavioral ecology and marine ornithology in the School of Biological Sciences at the University of California-Irvine. He was a member of the ecology panel of the NRC Committee to Review the Outer Continental Shelf Environmental Studies Program (1987-1992). His research is in the areas of ecology and reproductive biology of seabirds, coloniality and reproductive success, biological oceanography of seabirds, habitat selection, and foraging behavior. He received a BA in 1965 and PhD (biology) in 1971 from Harvard University.

Robert R. Jordan is State Geologist and Director of the Delaware Geological Survey (1969-) and a professor of geology at the University of Delaware (1988-). He is a member of the NRC U.S. National Committee on Geology (1990-1993). He was a member of the NAS Committee on Offshore Energy Technology (1978-1980). He was a governor's representative on the Outer Continental Shelf Research Management Advisory Board of the Department of the Interior (DOI) (1974-1977). Since 1985, he has served as Delaware's representative on Outer Continental Shelf Policy Committee, and is currently chair (1992-1994). He is a member of the Association of American Petroleum Geologists, the Association of American State geologists (president 1983-1984), the American Institute of Professional Geologists, and a fellow of the Geological Society of America. His research includes sedimentary petrology, stratigraphy, geology of the Atlantic Coastal Plain, micropaleontology, and ground water supplies. He received a BA in 1958 from Hunter College, an MA in 1962 and a PhD (geology) in 1964 from Bryn Mawr College.

Stephen J. Langdon is professor (1976-) and chair of the Department of Anthropology, University of Alaska, Anchorage (1987-). He is a member of the American Anthropological Association, the Society for Applied Anthropology, Council on Education and Anthropology, the Alaska Anthropological Association, the American Ethnological Society, the Society for Economic Anthropology, and the Canadian Ethnological Society. His research includes ecological, economic, and educational anthropology, maritime societies, policy, and theory of the northwest coast of Alaska. He received a BA in 1970, MA in 1972, and a PhD (anthropology) in 1977 from Stanford University.

June Lindstedt-Siva is a manager of environmental protection with the Atlantic Richfield Company (1981-). She was a member of the ecology panel of the NRC Committee to Review the Outer Continental Shelf Environmental Studies Program (1987-1992). She is a member of the Society of Petroleum Industrial Biologists (founder and former president), the Marine Technology Society, AIBS, AAAS, and Sigma Xi. Her research is in chemoreception in aquatic animals, especially chemical control of feeding behavior of sea anemones, the effects of oil on marine organisms, oil spill response planning, oil spill cleanup and control, environmental planning in industry, and implementing planning during the early stages of project development. She received her BA in 1963, an MS in 1967, and a PhD (biology) in 1971 from the University of Southern California.

H. Joseph Niebauer is a professor of marine science at the University of Alaska (1982-). He was a member of the physical oceanography panel of the NRC Committee to Review the Outer Continental Shelf Environmental Studies Program (1987-1990). His research includes physical oceanography, dynamics and numerical modeling of shelf, shelf break and frontal systems, the air-sea-ice interaction; and biophysical marine interactions. He received a BS in 1967 and PhD (oceanography/limnology) in 1976 from University of Wisconsin, Madison.

James J. Opaluch is a professor of resource economics at the University of Rhode Island. He was a member of the socioeconomics panel of the NRC Committee to Review the Outer Continental Shelf Environmental Studies Program (1988-1992). His research is in resource economics, damage assessment, and the valuation of natural resources. He received a BA in 1975 from University of Rhode Island, and an MA in 1977 and a PhD (agricultural and resource economics) in 1981 from University of California, Berkeley.

Robert T. Paine is a professor of zoology at the University of Washington (1971-). He is a member of the National Academy of Sciences, and was a member of the ecology panel of the NRC Committee to Review the Outer Continental Shelf Environmental Studies Program (1987-1992). He is a member of the Ecological Society of America (president 1979-1980), the American Society of Naturalists and the American Society of Limnologists and Oceanographers. His research is in algal ecology, and prey-predator relationships and invertebrate natural history. He received a BA in 1954 from Harvard University, a MS in 1959 and a PhD (zoology) in 1961 from the University of Michigan.

Richard M. Procter is a geological consultant and senior partner of Pras Consultants, Calgary, Canada since 1991. From 1980-1991, he was executive director for the Petroleum Resource Appraisal Secretariat of the Department of Energy, Mines and Resource of Canada. He has worked for the Geological Survey of Canada (GSC) since 1960 in a variety of research and managerial positions and was central to the development of a petroleum resource evaluation program within the GSC. He is a member of the American Association of Petroleum Geologists. He received a BS (general) in 1952, BS (geology) in 1953, and MS in 1957 from the University of Manitoba, and a PhD in 1960 from the University of Kansas.

Wilford F. Weeks is professor of geophysics and chief scientist at Alaska Synthetic Aperture Radar Facility, University of Alaska, Fairbanks. He is a member of the National Academy of Engineering (1979-) and a member of the NRC Polar Research Board (1975-1977, 1989-). He was a member of the NRC Committee on Polar Research (1971-1975); chair of the NRC Panel of Glaciology of the Committee on Polar Research (1971-1975, member 1966-1971); chair of the NRC Committee on Glaciology of the Polar Research Board (1975-1977); a member of the NRC Panel on Polar Ocean Engineering (1977-1978); member of the Environmental Criteria Working Group of the NRC Committee on Offshore Energy Technology (1978-1979); a member of the NRC Panel on Ice Mechanics (1980-1981); a member of the NRC Committee on Earth Sciences (1982-1984); a member of the NRC Arctic Remote Sensing Group (1983-1984); and a member of the NRC Polar Oceans Panel (1984-1987). He is a fellow of the Arctic Institute of North America, the Geological Society of America, and the American Geophysical Union; he is a member of the International Glaciology Society. His research is in the area of geophysics of sea, lake, and river ice. He received a BS in 1951 and MS in 1953 from University of Illinois, and a PhD (geology) in 1956 from University of Chicago.

Clinton Winant is a professor of oceanography at Scripps Institute of Oceanography, Coastal Resources Center. He was a member of the physical oceanography panel of the NRC Committee to Review the Outer Continental Shelf Environmental Studies Program (1989-1990). His research includes fluid mechanics. He received a BS in 1966 and MS in 1967 from Massachusetts Institute of Technology, and a PhD (aerospace engineering) in 1972 from the University of Southern California.